MW01127297

Principles of Controlled Maintenance Management

PRINCIPLES OF CONTROLLED MAINTENANCE MANAGEMENT

P. DALE JOHNSON

Published by
THE FAIRMONT PRESS, INC.
700 Indian Trail
Lilburn, GA 30047

Library of Congress Cataloging-in-Publication Data

Johnson, P. Dale.
Principles of controlled maintenance management/P. Dale Johnson.
p. cm.
ISBN 0-88173-354-7 (print version) -- ISBN 0-88173-396-2 (electronic version)
Plant maintenance -- Management. I. Title.

TS192.J67 2001
658.2'02--dc21

2001050115

Principles of controlled maintenance management/P. Dale Johnson.

Published by The Fairmont Press, Inc.
700 Indian Trail
Lilburn, GA 30047

Printed in the United States of America

10 9 8 7 6 5 4 3 2 1

0-13-008278-3 PH
0-88173-354-7 FP

Distributed by Prentice Hall PTR
Prentice-Hall, Inc.
A Simon & Schuster Company
Upper Saddle River, NJ 07458

Prentice-Hall International (UK) Limited, London
Prentice-Hall of Australia Pty. Limited, Sydney
Prentice-Hall Canada Inc., Toronto
Prentice-Hall Hispanoamericana, S.A., Mexico
Prentice-Hall of India Private Limited, New Delhi
Prentice-Hall of Japan, Inc., Tokyo
Simon & Schuster Asia Pte. Ltd., Singapore
Editora Prentice-Hall do Brasil, Ltda., Rio de Janeiro

Table of Contents

Chapter 1

Introduction

In the early stages of industrial development, maintenance practices were simple, primarily of the housekeeping and breakdown types. However, as the complexity of facilities, equipment and systems increased, so did the problems and expenses involved in maintenance operations. It became increasingly apparent that improvement of maintenance management practices and procedures was essential to achieve efficiency and effectiveness of the maintenance operations.

Just keeping a plant and its equipment sanitary, presentable and operating properly today is a carefully scheduled and well-managed operation that is supported by the finest equipment and products. Millions of dollars are spent annually in developing and improving maintenance tools and materials; this, in turn, saves many more millions of dollars in plant investment.

Breakdown maintenance is the term that most nearly describes the absence of planned and scheduled inspections and preventive maintenance. It needs little in the way of organization, but considerable in the variety of skills necessary for the repairman. For such work, no special supervisory planning is needed for repairing inoperative machinery and deteriorated structures. The maintenance personnel must, however, work hard just coping with emergencies and breakdowns.

A good maintenance management system does not have to be complicated. In fact, the best and most successful system is one that is simple, workable and gives the desired results. The more complex a maintenance management system is, the more chance there is that it will fail or not achieve the desired results and potential cost savings.

The system described in this book is simple but comprehensive and flexible. It is adaptable to any type of plant or facility in any location. It is also adaptable to any size of maintenance department from the smallest to the largest.

This book will show you how to develop, implement and manage your own controlled maintenance management system. It will also teach you how to improve your company's profits. You may ask how maintenance, a service function supporting production, can improve the company's profits because the profits come from the sale of the produced product. Let's take an example. Say that your maintenance is costing $20,000,000 per year for labor and materials. If you can reduce those costs by 10% through good maintenance management practices, you have saved $2,000,000 which is an added profit for your company and you, therefore, are "managing maintenance for profit."

The first step in developing and implementing a controlled maintenance management system is to obtain the concurrence of your top management.

The second step is to inform the maintenance personnel. They will be doing the actual work so you must have their cooperation. You should explain the system to them, how it will function, how they will benefit by the work being planned, the reduced breakdowns and reduced emergencies. You may encounter resistance to change by some of the people. This is not unusual. Some people feel comfortable with the way they have been doing their work for many years and are afraid of change. One way to win them over to the change is to compare the driving and riding comfort of the 1920s automobiles and the present day automobiles. They may be comfortable with the breakdown maintenance method of the 1920s but the modern method is much more comfortable and easier.

The third step is to decide where you will start. Should you start with one area, one group of machines or the whole plant? This can be determined by the size of operation that you have. If you have a large plant, it is often easier, and sometimes advisable, to start with one area. By starting with one area, you can start the system on a relatively small scale, make any needed changes and then progress through the plant.

The fourth step is to make an inventory of the equipment, systems and facilities. The equipment inventory should list the type of equipment (pump, punch press, etc.), manufacturer, model and serial number, utility requirements (air, water, voltage), location (Building 23, southeast corner) and, if possible, the purchase date, cost and depreciation rate. It may seem strange to have the purchase date, cost and depreciation rate information in a maintenance record but it is important. Your analysis of the item's history file will assist you in determining when an equipment item is reaching the end of its economic life. This becomes your supporting data for your request for a replacement.

It is suggested that you start the system with a

manual record system. This will allow you to easily make changes as you progress.

You will then have a system that is customized to your facility, equipment and needs. After the manual system is working satisfactorily, you may want to computerize it. The computer will make it easier to enter and retrieve information. Do not design the computerized system as a rigid system, it must remain flexible for future changes.

DEFINITIONS

Facility

A separate building, structure or real property improvement that is built, installed or established to serve a particular purpose.

Equipment

The machinery or capital assets, other than buildings or structures, designed, manufactured, installed or established for specific use in the operation of an activity, facility, system or utility.

System

A combination of buildings, structures and/or equipment built or installed for the provision, generation and/or distribution of essential services, for example, water, compressed air, electrical, sewage, etc.

Utility

See "System" above.

Maintenance

The recurring day-to-day, periodic or scheduled work required to preserve or restore facilities, systems and equipment to continually meet or perform according to their designed functions.

Backlog

The accumulation of a reserve of planned and estimated maintenance work for future accomplishment to ensure a continuous workload for the maintenance work force.

Development and Implementation of a Controlled Maintenance Management (CMM) System

The successful development and implementation of the CMM system consists of five phases:

Phase I

1. Inventory of facilities, systems and equipment. The inventory will include:

   Facility—type and year of construction, size, identification.

   Equipment—type (pump, punch press, etc.), manufacturer, model, serial number, utility requirements (air, volts, water), identification number, location.

2. Analyze current maintenance management methods and philosophy.

3. Analyze current maintenance recording and tracking procedures.

4. Procure manual system to record maintenance performed.

Phase II

I. Design work request and work order forms.

2. Establish work request and work order procedure flow.

3. Establish work order priority codes.

4. Establish requisition and purchase order flow.

Phase III

1. Train personnel on historical record keeping.

2. Train personnel on work request and work order procedures.

3. Develop preventive maintenance inspection (PMI) points and schedules.

Phase IV

1. Determine starting point of PMIs (area, type of equipment, group of like machines, etc.) and implement PMIs.

Phase V

1. Issue work orders based on problems found during PMIs.

2. Evaluate PMI schedules and make adjustments (decrease/increase) as indicated by problems found.

Phase VI

1. Computerize the system when the manual system is operating satisfactorily.

Chapter 2

Principles of Controlled Maintenance Management

Basic Concepts and Objectives

Controlled maintenance management is an organized, systematic and controlled approach to maintenance and its management through the effective application of the following basic principles:

ORGANIZATION
INVENTORY
CONTINUOUS INSPECTION
PLANNING
SCHEDULING
MANAGEMENT ANALYSIS

The system consists of organizing, planning, supervision, coordination and control of the functions necessary to ensure that a plant can continually perform its designed function in an economical manner within budgeted limits. The system ensures that the plant will obtain the most efficient utilization of labor, equipment and materials in achieving this goal. The continuing objective of the control effort is to increase productivity, achieve savings and ensure that the designated level or standard of maintenance is achieved.

Significance of Controlled Maintenance Management

Purposes

To fully understand the significance of the controlled maintenance management system and the benefits to be derived from its implementation and successful administration, it is essential that the basic purposes be clearly established. They are:

a. To ensure that the maintenance level or standard of buildings, grounds, facilities, systems, utilities and equipment is maintained to meet their designed functional requirements;

b. To perform scheduled routine maintenance thereby reducing, extending or eliminating the requirement for major repairs and/or costly replacement of facilities and equipment;

c. To provide positive control of the maintenance labor force through competent supervision;

d. To improve shop efficiency by balancing the shop

work load with the work force by means of proper work scheduling;

e. To provide a stable work force through projected shop loading techniques based on known work backlogs;

f. To control maintenance costs by the continuous evaluation of actual versus estimated expenditures and work performance through the use of management analysis reports;

g. To provide a centralized facility and equipment records and costing system.

Requirements

In order to achieve the degree of effectiveness and efficiency necessary to implement the system, the cooperation of all levels of management, supervisors and craft personnel is necessary.

Therefore, it is a primary requirement that all levels of personnel be familiar with the system's functions and, equally important, that each individual be indoctrinated to realize and appreciate the value of the contribution he/she can make toward the accomplishment of the system objectives.

Benefits

Successful and complete implementation and administration of the system results in increased economy and greater reliability in maintenance operations, improved morale of maintenance personnel, increased productivity and the provision of data to support budget requests. In addition, it manifests the development and application of improved technical data and maintenance methods.

Basic Elements of Maintenance Control

Implementation of the controlled maintenance management system, and the extent of efficiency and effectiveness it attains, is based on the application of ten basic control elements. Each element adds to the measure of control achieved and represents an essential and integral part of maintenance management. The control elements are procedures utilized to achieve the desired objectives. They provide the necessary administrative and supervisory controls required in the expenditure of labor, materials and equipment for the accomplishment of maintenance.

Inventory and History Files

A detailed and complete inventory of all buildings, facilities, utilities, systems and equipment must be made to determine the scope of the maintenance effort required, the condition of the facilities and equipment and the types. The completed inventory provides the necessary data to establish the facility and equipment files.

Maintenance Standards

Maintenance standards establish the acceptable condition of buildings, grounds, facilities, utilities, systems and equipment and determine the level of maintenance allowed to maintain the standard. The standards are determined by the designed function, amortization, type of construction, age, purpose, present and future requirements and replacement cost. Knowledge and application of the standards permit decisions to be made concerning the extent, frequency and necessity of inspections and maintenance of facilities, systems and equipment. They also provide a source of comparison between the actual condition and the acceptable standard for a particular unit.

Continuous Inspection

Continuous inspection is the planned and scheduled inspection of facilities and equipment to locate sub-standard conditions and deficiencies in the early stages of development and to initiate corrective action. Continuous inspection is the major source of work generation for the maintenance forces. It provides a stable flow of work, builds an essential work backlog and permits better planning for utilization of funds, manpower, materials and equipment.

Continuous inspection consists of two types of inspection. Preventive maintenance inspections (PMI) are performed by the individual shop forces. PMI consists of inspection, lubrication, minor adjustments and repairs of facilities, systems and equipment.

Operator inspection (OI) consists of examination, lubrication and minor adjustments of equipment, systems and utilities to which specific operators are assigned.

Work Classification

Work classification is the procedure that channels and prescribes the handling and management of each type of work from inception to completion and review. Factors determining the classification include the probable duration of the work, urgency, nature and the purpose of the work. There are five basic work classifications:

Emergency work
Service work
Minor work
Specific work
Standing work

Work Input control

Work input control is the classification, identification and recording of work generated by continuous inspection, customer requests and other sources or methods. A central work reception center is established to process the input of work. Through the proper application of procedures and governing criteria, unnecessary requests for estimates and maintenance work will be either held to a minimum or eliminated. The work reception center processes work input to assure proper funding and approval of the work, simultaneously maintaining positive control of the transactions involved.

Planning

Planning and estimating responsibilities comprise a major portion of the functional controls included in the controlled maintenance management system. Planning and estimating is responsible for the screening of work, preparation of a manpower/material/equipment cost estimate and planning the sequence of operations in accomplishing the work.

Material Coordination

Material coordination is the application of necessary controls to material procurement, monitoring, storage and delivery functions to ensure that the required materials are available and ready for issue at the time work is started. It is essential to the orderly and systematic progress of the work and is one of the major determining factors involved in scheduling.

Scheduling

Scheduling is the application of shop loading techniques to ensure that the work load is balanced with the work force and that an adequate reserve of work (backlog) exists to assure a constant and steady flow of work to the shops. Scheduling stabilizes the maintenance forces and permits the projection of manpower forecasts to meet future requirements. As a carefully prepared advance plan of action it coordinates the various shops to achieve the maximum efficiency of work performance. Scheduling consists of the following:

- Projected shop loading where work is scheduled by month for a period of three or more months in advance based on projected manpower and material availability.

- Individual shop scheduling by shop supervision provides the daily assignment of shop personnel and covers a one week period.

Work Performance and Evaluation

Of primary importance in determining the effectiveness of controlled maintenance management is work performance and evaluation. Comparison of estimated versus actual labor, material and equipment expenditures, their evaluation and analysis, assists in locating areas of deficiencies within the system. The deficiencies may exist in planning and estimating, material coordination, work scheduling, actual shop performance or a combination of all or part of them. Pinpointing the trouble area permits initiation of corrective action. Work performance evaluation also determines the cost effectiveness of the maintenance department.

Reports

Management analysis reports are a tool to provide management with the facts needed to evaluate the distribution and utilization of funds and manpower within the maintenance department and provide necessary information for work performance evaluation. Through the use of analysis reports, management is kept informed on work performance, work variations, the ratio of material and labor expenditures, manpower utilization, the need for manpower increases or decreases, work backlogs and cost effectiveness and can isolate potential trouble areas for corrective action.

Organization

The maintenance department of a company is the in-house provider of essential services. It is essential that the maintenance department is always involved in plant layout planning to assure that equipment is not located in a position where it cannot be maintained.

The maintenance department has overall responsibility for the administration, supervision and operation of the maintenance shops and plant operations and for performing maintenance and repair of facilities, systems and equipment. Plant operations will include steam, power generating, air compressors, pumping and water plants.

The effective implementation of the controlled maintenance management system and achievement of the full value of the system requires the establishment of a proper organizational pattern. Clearly defined lines of authority and responsibility must be established and the functions of each component of the department clearly understood if maximum performance and efficiency is to be attained.

Although a rigid organizational structure is not practical in many cases because of varying geographic and/or technical circumstances, the basic organization of the controlled maintenance management system is adaptable to any location and any size of maintenance department. It provides a sound basis for organization and administration of the maintenance department.

The entire effort of the maintenance control function is directed toward controlling the maintenance functions and workforce. Maintenance control accomplishes the control/management function for the maintenance department.

The primary responsibilities of maintenance control include the content of the preventive maintenance inspection (PMI) lists and frequency of inspections; facility and equipment historical files; estimating (labor and materials) and planning all work; development of the maintenance backlog; material procurement; justification of projected major maintenance, repair and alteration projects; preparation of manpower forecasts and management reports of work performance.

CHAPTER 3

INVENTORY AND HISTORY FILES

An inventory of all items of facilities and equipment is the foundation of the controlled maintenance management system. Each facility and equipment item must have its own inventory or history file. It should list all pertinent information such as nomenclature, type, manufacturer, model number, serial number, identification (Pump No. 1, P-1), purchase date, installation cost, utility requirements, location, etc.

INVENTORY AND HISTORY FILES

The history files provide a central source of reference data. These data can be retrieved readily and include all completed maintenance and repair including labor and material costs and all actions accomplished on a particular facility or equipment item. Information that can be obtained from the file include: Repetitive adjustment of a particular machine or repetitive repair of a particular part on a machine.

A determination of when an item is reaching the end of its economical useful life. The determination will be substantiated by accurate data showing the annual and increasing cost of repair. If there is a question of rebuilding the item or replacing it, compare the present value plus the cost of rebuilding versus the cost of a new replacement.

Example:

Present value of machine	$ 3,500
Cost of rebuilding	45,000
Estimated cost of repairs next year	7,500
	$56,000

Cost of new machine	$46,000
Cost of installation	3,500
Estimated cost of repairs first year	500
	$50,000

$56,000 – $50,000 = $6,000 saved in the first year by replacing the old machine.

Additional information that will be obtained:

- Accurate data on the cost of maintaining various types of equipment and facilities under various conditions of use, climatic conditions, etc.
- A means to evaluate various types of construction and maintenance practices.

- The frequency of inspections and the deficiencies found.

Scope and Type of Information

Data accumulated in the history files should be limited to useful facts and statistics of a long range nature. Information that should be included are dates and types of repair, labor and material costs and running totals of costs. All entries should be by fiscal year. Completed work orders, with labor and material costs, will provide the required posting data for the permanent record.

When an electric motor or other item of equipment is replaced, subsequently repaired and put into the stock room for later use, the item's history file should show this information. The new location will be shown when the item is returned to service.

As-built Drawings

This is a very good time to get your as-built drawings up to date. If the facility is more than a few years old it is quite likely that the utility distribution systems, i.e., water, steam, electrical, etc., have been modified and no one recorded the changes. This can lead to dangerous situations. Does everyone know where the shut off valve is in case of a rupture in a steam, air or water line? Or where the disconnect switch is for a particular electrical circuit? Even new facilities have errors in the as-builts. Any changes in a utility system should be recorded immediately on the appropriate drawing with the date of the change.

Updating the as-built drawings should be performed at the same time that the physical inventory is made. All valves, drains, traps, disconnects, gauges, and equipment should be given an identifier number, e.g., pump - P43, and noted on the as-built drawing.

If you are making the inventory by area, the as-built drawings can be made by area. If you make the inventory by system, the as-built drawings should be by system. The various systems as-builts should be consolidated for easy reference when the updating is completed. Setting up the utility systems into major systems and subsystems will make the tasks much easier for the maintenance personnel when they are reviewing the as-builts. You may wish to put the as-builts on a computer for easier review and more permanency.

Another suggestion that maintenance personnel like is to develop maintenance manuals by area. Each manual will have schematics of the systems and subsystems. The schematics will show the location of each valve, gauge, etc., with the identifier number. The manual will contain the manufacturers' parts manuals with cutaway drawings with part numbers. The maintenance technician then knows the exact part number to draw from supply. This eliminates the guess work when replacing parts.

Chapter 4

Continuous Inspection

The purpose of continuous inspection is to find and identify deficiencies in the plant components and to initiate the corrective actions necessary to return the facilities, systems and equipment to a reliable condition. Inspections, with the needed corrections, will reduce the number of breakdowns and costs of repairing. They provide a constant flow of work to the maintenance forces and allows proper planning for the utilization of labor and material resources through the advance planning of work.

Types of Inspection

Continuous inspection encompasses two distinct types of inspection, preventive maintenance inspection (PMI) and operator inspection (OI).

PMI consists of examination, lubrication, minor adjustments and repairs of systems, utilities and equipment. It is particularly applicable to the unattended portions of such facilities as:

Water supply, treatment and distribution systems;

Electric distributions systems;

Overhead cranes;

Heating, ventilating, refrigeration and air conditioning systems;

Food preparation, service and dish-washing equipment;

Fire detection and suppression systems and equipment.

Determining Inspection Requirements and Frequencies

It is necessary that a determination be made concerning the extent of inspection and the type of inspection required.

The determined maintenance standard (level of reliability), local climatic conditions and production requirements combine to establish the extent of inspection applicable to a particular item of the plant. After careful evaluation of the above conditional factors, the specific inspection checklists can be assigned and the inspection frequency for each determined. Close coordination between the maintenance and production departments is essential during the initial stage of the program.

When the extent of inspection has been established, the next step is to determine the proper type of inspection required for its accomplishment. Whether the inspection should be PMI or OI is generally self-evident. The periodic and systematic inspections performed by qualified inspectors should be the major source of work generation for the maintenance department and are the only reliable means to ensure that the determined reliability level for the plant is achieved.

Preventive maintenance inspection (PMI)

The purpose of PMI is to assure the efficient operation of facilities, utilities, systems and items of equipment by providing for their systematic examination and the correction of minor deficiencies before they expand into projects requiring major repair or replacement.

Development of Inspection Checklists

An inspection checklist must be developed for each item that will receive PMI. An appropriate PMI checklist must be carried by the inspector when making the inspections to assure that nothing is forgotten or overlooked. It is recommended that you develop your own inspection checklists. You should start with the manufacturers' recommendations adapted to your particular usage and climatic conditions. Manufacturers' recommendations are general and not for any particular location. Next, ask the equipment operators and the maintenance personnel for their suggestions. You will now have inspection checklists for your particular usage.

Extent of Inspection

It is necessary that a determination be made concerning the extent of inspection. An easy way to determine the extent of PMI for an item is to follow these criteria.

What effect will a failure of the item have on the system? If its failure will disrupt a vital system, it should have frequent PMI. If its failure will cause only a minor inconvenience, it will require less frequent PMI. The replacement cost of an item will also have a bearing on the frequency of inspection. An expensive item of equipment, for economic reasons, will require more frequent PMI.

In some cases, replacement of an item may be more economical than PMI.

PMIs are performed on a scheduled basis by the maintenance shop personnel. The inspector should carry the tools and equipment that will be required to accomplish the inspection. The actual time allowed for a particular inspection will vary with the type of item inspected.

The inspector may not be enthusiastic about carrying a lot of papers when making PMIs and may work to assure that PMI does not succeed. A way to avoid the problem is to equip the inspector with a hand held computer. The computer should be downloaded at the end of each work day.

Overhaul of vital equipment or utility systems should be planned and scheduled to minimize interruptions of service. A scheduled overhaul is not a part of PMI but is a planning responsibility of the supervisor of the maintenance department. The schedule of PMI should be coordinated with the overhaul schedule to obtain the maximum benefit of both procedures.

The application of PMI to such items as automatic door closers, door latches, window locks and water coolers with sealed refrigeration units is not considered necessary. In some cases, replacement of an item may be more economical than PMI. The determination of items to receive PMI, after the vital operating equipment is selected, must be made by the maintenance department. Environmental conditions, age of the equipment, usage and character of usage vary from plant to plant. An analysis of trouble calls for nature, frequency and cost of service will provide a basis for a decision to apply or not apply PMI. The analysis can also indicate a probable frequency for inspection if it is decided to apply PMI. The annual cost of inspection and repair as compared to the replacement cost of the unit should be an important consideration. In addition, an analysis of the various specific services required at periodic intervals should be made. It is possible that these services, or some of them, can be performed under specific work orders issued at predetermined intervals, thus eliminating the need for PMI or increasing the time between inspections.

Inspection Record System

It is essential that a system of inspection records be established and maintained so effective and continuous inspections can be controlled with a minimum of time, money and manpower. It should be clearly understood that the purpose of the inspection record system is to do the job in an easy and economical manner. Only the forms and records that affect these economies should be used. Records should not be generated if there is no use for them.

Scheduling of Inspections

Economical performance of continuous inspection requires a schedule of inspections based on the inventory of facilities and equipment items. Factors listed in the following paragraphs should be considered in the preparation of the inspection schedules.

Frequency of Inspections

The frequency of inspections is a very important element of continuous inspection. Too frequent inspection can be a waste of manpower and can lead to over-maintenance while infrequent inspection can result in under-maintenance and large repair or replacement projects.

Frequencies of inspection and test should be based on local climatic conditions, age of the item, use including severity of use and other local factors considered pertinent.

The following is a simple guide to help determine the frequency of inspection. Ask the following questions:

- Will failure of an item endanger life and/or property?
- Will failure result in pollution control problems?
- Will failure of the item interfere with an essential operation of the facility?
- Does the item have a high cost or a long lead time for replacement?

If all of the questions are answered YES, then the item must have frequent inspection and the very best plan of maintenance that can be given.

If the answers to the questions are a combination of YES and NO, the item would fall into a second group which requires some attention but not to the extent of the first group. For example, failure of a ventilating system for a paint booth might not interfere with an essential operation but could endanger life or property for various reasons. The faulty condition that usually precedes a breakdown could cause a fire or explosion. Failure of the system to operate could endanger the health of personnel.

If the answers to all of the questions are NO, the item needs little or no preplanned maintenance. An example of this third group is an exhaust air system for showers and toilets. Failure of the system could cause some inconvenience but would not interfere with an essential operation, would not endanger life or property and does not usually have a high replacement cost.

Basis of Inspection Frequency

Inspection frequency may be established on a calendar basis such as daily, weekly, bi-weekly, monthly, quarterly, semi-annually or annually; on a seasonal basis such as twice a heating season; or on an operational basis such as 1,000 hours or 2,000 hours.

Maximum Frequency

Inspection of items should not be scheduled more frequently than required to assure normal operating efficiency in conformance to the applicable maintenance standards. In general, no item should be inspected less frequently than once a year. Exceptions to the requirement for annual inspections may be made for towers and similar items where it is operationally impractical to make internal or external inspection.

Replacement of Parts Upon Breakdown

On certain types of low-cost or inaccessible items, replacement of the item at the time of breakdown may be more economical than continuous inspection. When it is determined that replacement is economical, periodic inspections are not warranted. Examples of items that may be more economical to replace are household type sealed refrigerator units, fractional horsepower motors and automatic door closures. When such items are part of vital operating equipment, the items should be stocked or be readily available to permit replacement at the time of breakdown.

Adjustment in Frequency

As PMI becomes established in routine, it may be found that no deficiencies are found in various items. When deficiencies do not occur, consideration should be given to scheduling less frequent inspections. On the other hand, when the same deficiency is reported after each inspection of a particular item, inspections should be more frequent. Frequency of inspection should be such that emergency calls and service work are kept to a realistic minimum. One of the functions of the maintenance department is to analyze the maintenance work generated by emergency calls and work and, if advisable, to adjust the inspection frequency.

Inspection Frequency Tables

The tables of inspection frequencies provide suggested starting frequencies of inspection to maintain facilities, equipment and systems at or near the maintenance standard. The frequencies may require modification because of age, usage and condition, weather and other conditions.

Some Scheduling Criteria

Availability of Facilities and Equipment

The availability of facilities or items of equipment, the production requirements of a plant and the inconveniences that may be created for the operators of equipment or the occupants of buildings should be considered in establishing the inspection schedules. It may be necessary that some items be inspected outside of normal working hours.

Weather Conditions

Some deficiencies can be more easily found in inclement weather than in fair weather. This is particularly applicable to deficiencies in storm drainage and to ponding on pavements. It may be desirable to inspect roofs at a time such that necessary repairs may be accomplished prior to the start of the rainy season. Where weather conditions follow a consistent cycle, inspection schedules can be designed for them.

PREPARATION OF INITIAL INSPECTION SCHEDULES

An inspection form must be developed for each item of equipment and facility. The purpose of the form is to assure uniform reporting of all items of similar types of equipment. A periodic review of the inspection forms may reveal that particular items from the same manufacturer have the same discrepancies. This is a signal for an investigation to determine the cause.

SCHEDULING

One purpose of PMI, in addition to locating deficiencies in the early stages of development and correcting them, is to provide a more constant flow of work to the maintenance department. It is necessary to make several determinations before preparing an inspection schedule. Is the activity to be divided, for inspection purposes, into areas? If the activity is divided into areas, should the inspection records be kept by those areas? Or should the inspections be by systems? Or should they be by like types of equipment? Only the people involved can make these decisions.

The following procedures provide details on the actions for scheduling inspections.

1. Prepare an inspection checklist for each type of equipment or facility. Use a separate list for each type of inspection (electrical, mechanical, structural) involved.

2. In scheduling inspections, it is not necessary or advisable to predetermine the period in which the item is to be inspected closer than 1/2 month for items on monthly and longer periods. The inspector is thus given a workload for 1/2 month and may arrange his performance to his best advantage.

3. Having determined a schedule and made any desired adjustments, the period of the first inspection should be written in the "Date Inspected" column of the inspection record. The period can be shown as 1/2 Oct. or 2/2 Oct., meaning the first half or second half of October.

4. The inspection checklists should now be sorted by type of inspection and placed in a Kardex or other type of file.

5. Scheduling for PMI can now be undertaken, following in general the same procedure given in paragraphs 2 through 4 above.

Responsibility of Inspectors

The success of an inspection program depends upon the quality and experience of the personnel selected to make the inspection. Inspections should be made by qualified personnel from the appropriate maintenance shops.

The inspectors should be conscientious and extremely careful in their observations. They should take sufficient time to make thorough examinations; conclusions should be reached only on the basis of actual observation and analysis. When a thorough examination is not possible, inspectors should make a report to this effect. Inspectors should consider defects carefully to determine the relation to, or influence on, the safety of the equipment. They should question responsible employees as to any past or present problems, including pertinent facts. Also, the inspectors should determine what repairs have been made and whether such repairs were made properly. The inspectors should make every effort to determine the true cause of a deficiency. An apparent cause is not always the true cause and may only be a secondary reaction or manifestation of the true cause. To achieve economy in maintenance, the true cause of a deficiency must be corrected. The inspectors should make a general observation of the condition of the facilities and equipment as a guide to forming an opinion of the general care of the facilities and equipment.

Safety

Appropriate protective equipment, such as safety glasses, hard hats, gloves, respirators, safety belts, lockouts and the like should be provided to each inspector when applicable. The inspectors should observe the proper safety precautions in the conduct of inspections and test, not only to avoid hazards to themselves but also to avoid creating hazards for others. Some inspections will require the presence of a helper to minimize any danger. Not only must inspection personnel observe all safety precautions as a matter of safety for themselves, but also to provide an example to other personnel working in the area.

Indoctrination

There may be situations in which inspectors should be given specific indoctrination. Some inspectors may require more indoctrination than others because of the type of inspection conducted. The safety supervisor should indoctrinate the inspectors in the maintenance and use of safety equipment.

Inspectors should carefully and adequately describe the deficiencies and the suggested corrections. Complete reports from the inspectors will help in estimating and planning the repairs and minimize trips to the site to obtain missing information.

General

Inspectors should be fully instructed concerning the duties expected of them and they should be held responsible for conducting a thorough and workmanlike inspection or all units and facilities assigned to them. Inspectors must be aware at all times of the applicable maintenance standards and of any current austerity directives to avoid the performance of unnecessary work and to avoid making recommendations for unnecessary repairs. Inspectors should also be counseled to request technical assistance when they doubt the sufficiency of their judgment in particular cases. For example, the inspector may determine that a roof truss requires repair because there is considerable sag in the truss and some of the joints show evidence of movement but he is not sure of the proper correction needed.

Action During Inspection

Use of the Inspection Checklist

The inspector examines or tests the item for the check points listed on the inspection checklist for the inspection. Using the proper code, he indicates the condition of each check point. At the conclusion of the inspection, he enters the number of inspection hours, to tenths of an hour, indicates whether a deficiency report was made, dates and initials the inspection report.

Breakdown

If the inspector finds a deficiency that requires immediate correction to prevent loss or damage to property, to restore essential services that have been disrupted by a breakdown of utilities or to eliminate hazards to personnel or equipment, he should report the situation to his supervisor immediately.

Deficiencies

When the inspector finds a minor deficiency that he can correct in a few minutes with the tools that he carries, he should correct it. If the deficiency is not critical and the time to correct it exceeds the inspector's allotted time, the inspector should describe it on a deficiency report. The true cause of the deficiency should be recorded and a sketch made with measurements, if it will describe the required work more clearly. The inspector should make as complete a description of the deficiency as he can in order to minimize other visits to the site. Failure to make a complete description will usually involve additional expense and, in some cases, become a nuisance to the occupants. Inspectors should not spend more than fifteen to twenty minutes inspecting each item of equipment. The deficiency reports are submitted to the maintenance office for estimating, planning and scheduling of the repair work order.

Return of the Inspection Checklists and Reports

At the end of each day or at the end of the inspection period, depending upon the number of deficiencies found, the inspector's reports and the corresponding lists are returned to the maintenance office.

Operator Inspection (OI)

The purpose of operator inspection is to assure efficient operation by providing systematic and continuous examination of equipment, systems, utilities and items of equipment to which a specific operator is assigned. Operator inspection relieves the maintenance department of some of the workload.

Scope

OI consists of examination, lubrication, minor adjustments and repairs of systems, utilities and equipment to which a specific operator is assigned, for example, a machine operator or a boiler plant operator. The inspection is performed by the operator during the normal hours of the operator's workday.

Method

The inspections are performed in accordance with the manufacturer's operating instructions and plant developed checklists. Deficiencies beyond the capacity or authority of the operator are reported to the immediate supervisor on a deficiency report. Breakdowns of equipment should be reported immediately.

CHAPTER 5

WORK CLASSIFICATION

Work classification is the procedure that channels and prescribes the handling and management of each type of work from its inception to its completion and review.

SCOPE

All work accomplished by the maintenance department should be classified into one of five categories. Determining factors for work classification include the type of funds involved, the probable duration and extent of the work, the urgency, nature and purpose of the work.

CATEGORIES OF WORK CLASSIFICATION

Controlled maintenance management has five classifications of work: emergency, service, minor, specific and standing work.

Emergency Work

Characteristics

The primary characteristic of emergency work is that it requires immediate action to accomplish any or all of the following purposes involving the plant:

(1) To prevent loss or damage to property,
(2) To restore essential services,
(3) To eliminate hazards to health or personnel.

The work is normally accomplished without an emergency work authorization (EWA). The EWA is written upon completion of the task. If, upon completion, the cost of labor and materials is found to exceed the established limitation for the EWA, it is changed to a minor or specific work order, as appropriate.

Source

Emergency work can be requested by anyone.

Control

The work requested is determined to meet the criteria established for emergency work. EWAs are usually limited to two man-days and an established material limitation.

Service Work

Characteristics

Service work is work that is relatively minor in scope, is not emergency work by nature and does not exceed the established man-hour and material dollar limitation. For example, a complaint may be received that a fan is noisy.

The problem may be that the drive belt is slipping and causing a squealing sound.

Source

Service work can be requested by anyone in writing, in person or by telephone.

Method of Authorization

Service work is authorized on a service work authorization. If the work will exceed the established limitation for service work authorization, the maintenance supervisor should be notified so the necessary steps can be taken to supersede it with a minor or a specific work order, as appropriate.

Control

If the work requested meets the criteria established for service work, it is not planned, estimated or scheduled. Minimal control is exercised on service work.

Minor Work

Characteristics

Minor work is work in excess of that authorized by service work authorization and less than that authorized by a specific work Order. It is planned and estimated with the degree of accuracy necessary to assure effective shop scheduling.

Source

Minor work can be generated by an emergency or service work authorization or a maintenance work request.

Method of Authorization

Minor work is authorized on a Maintenance work order in accordance with established approval authority.

Control

Minor work orders are planned and estimated. Generally minor work orders are used as fill-ins or supplemental work to employ labor not scheduled for specific work orders or other work, thus assuring a constant high level of labor utilization.

Specific Work

Specific work orders are issued to authorize the accomplishment of a specific amount of work. It is work in excess of that authorized by emergency or service work authorizations or a minor work order.

Source

Specific work can be generated by an emergency or service work authorization, preventive maintenance inspection report or a maintenance work request.

Method of Authorization

Specific work is authorized on a Maintenance work order in accordance with established approval authority.

Control

Specific work orders are planned, estimated and scheduled. Maximum control is exercised on specific work to include detailed planning, estimating, material coordination, the application of scheduling techniques and close supervision.

The minimum man-day requirements of work to be authorized by specific work orders must be carefully determined by an analysis that considers both the percentage of the work force to be controlled and the relative administration costs of processing individual work orders. A basic aim of controlled maintenance management is to establish adequate controls at a minimum cost.

Standing Work Orders

Characteristics

Standing work orders include all work that is highly repetitive in nature such as PMIs. This category covers both labor and material costs. PMIs should be grouped according to area, system, equipment, etc. for the work concerned. Standing work orders are utilized to consolidate costs and eliminate a large amount of paper work. Standing work orders are issued on a quarterly basis and are a means of accumulating cost data.

Source

Standing work orders are usually generated as a result of the maintenance and operational requirements of the plant. They are also generated by maintenance work requests to provide necessary services and support for other departments.

Method of Authorization

Standing work is authorized on a maintenance work order in accordance with established approval authority.

Control

Maximum control in terms of final estimating, planning and scheduling is not possible for standing work orders. Scheduling is accomplished on schedule boards only and, therefore, only moderate control is exercised.

Types of Standing Work

Estimated Standing Work Orders

This is highly repetitive work for which manpower requirements are constant and predictable over protracted periods of time. An exact description of the work to be accomplished, a specific frequency cycle and man-hour estimates are to be included in estimated standing work orders.

Unestimated Standing Work Orders

This is work for which manpower requirements cannot be estimated accurately. Work requirements should be analyzed carefully and unestimated standing work orders issued only after a clear determination that specific work orders or estimated standing work orders are not applicable.

Work Load Analysis

To effectively determine the most feasible and economical use of manpower in accomplishing minor and specific work, it is necessary to analyze the workload. This is accomplished by determining the number of work orders and the man-hours used, by man-hour groups, and converting these figures into percentages showing (a) the cumulative percentage of work orders and (b) the cumulative percentage of man-hours (see work order analysis, page 46).

The analysis should be based upon data for at least a three-month period. The recommended breakdown between minor and specific work is:

MINOR WORK		SPECIFIC WORK
25%	to	75%
30%	to	70%

The 70%-75% figure for specific work orders represents the amount of the work force which should be subjected to the controls provided by individual work order cost reporting procedures and scheduling techniques.

The objective of a work order analysis is to control the majority of the work force available. Therefore, the first point of reference in the analysis will be the figure that is nearest the 70%-75% man-hour level.

The work load analysis shows a typical work order analysis. The desirable range of cutoff between minor and specific work orders, based on the 70%-75% figure for specific work, would be 54 man-hours. In this category, only 37% of the work orders are involved and 74% of the man-hours are controlled. Therefore, specific work would be all work in excess of 54 man-hours and minor work would be all work that is 54 man-hours or less. However, the final decision as to what man-hour level is chosen must be your decision.

WORK ORDER ANALYSIS

MAN-HOURS	CUMULATIVE PERCENTAGE OF WORK ORDERS	CUMULATIVE PERCENTAGE OF MAN-HOURS
Over 0	100.0	100.0
Over 9	94.0	98.0
Over 18	76.0	95.0
Over 27	62.0	89.0
Over 36	53.0	84.0
Over 45	42.0	80.0
Over 54	37.0	74.0
Over 63	30.0	68.0
Over 72	26.0	66.0
Over 81	20.0	65.0

CHAPTER 6

WORK INPUT CONTROL

Work input control is the reception, classification, identification, recording, processing and control of all maintenance work including emergency and service work and the work generated by Continuous Inspection, maintenance work requests and other sources.

SCOPE

All work for maintenance department accomplishment, from its inception to its reviewed completion, should be processed in accordance with the suggested procedures and governing criteria set forth in this chapter.

WORK RECEPTION CONTROLS

To achieve the maximum control of work input, all work must be channeled through a central work reception center located in the maintenance department.

The work reception center accomplishes two related

but distinct functions: The processing of all emergency and service work requests and the processing of all other work.

I. Work reception center (trouble desk)

All emergency and service work is processed through the work reception center. The processed work falls into two categories: Work for which standing work orders have been established and funded and work chargeable to maintenance and operations accounts.

A. Source and Method of Request

1. Maintenance and Operations Support Work
   This work can be requested by anyone in writing, in person or by telephone.

B. Method of Authorization
Emergency and service work which does not exceed the man-hour/material cost limitations established is authorized for accomplishment on an emergency/service or minor work authorization. Emergency and service work is processed in the following manner.

1. The request is received at the work reception center and all essential information is written on the order. The information must be checked at the time of the request to assure accuracy, completeness and legitimacy.

   The requestor's name and phone number is always taken for reference purposes. In addition, when applicable, the name of the individual to contact and phone number should also be obtained.

2. An emergency, service or minor work authorization is prepared providing all of the information necessary for the shop forces to perform the work. The appropriate account number is assigned at this time.

3. The work reception center can approve emergency or service work involving maintenance funds.

C. Accomplishment of work
The emergency, service or minor work authorization is issued to the shop concerned. The shop supervisor will assure that the work is performed, complete the form and return it to the work reception center.

D. Records
The work reception center will, upon receipt of the original work authorization, destroy any suspense copies and record the original in the manner prescribed for the following.

1. Work performed on installed equipment will be recorded in the respective equipment history file.

2. Work performed on plant facilities, utilities or systems will be recorded in the appropriate file.

When the work requested exceeds the established limitations for emergency and service work, it should be processed in the manner prescribed for minor and specific work.

II. Work reception center

All work must be processed through the work reception center for record keeping purposes. Record keeping of work requests will eliminate issuing duplicate work orders and wasted time of maintenance personnel. It is a key element in maintaining control of the work force.

A. Source

The work can be generated by continuous inspection deficiency reports, maintenance work requests or other ways such as special engineering reports, recommendations, etc.

B. The work reception center should maintain a log book for recording work and controlling the transactions involved in its processing. The minimum requirements for a log book include columns and spaces for the following items.

1. A control or work identification number.
2. Date of receipt.
3. A brief description of the work.
4. The name of the requestor.
5. The initials of the Planner having estimating responsibility.
6. The number of estimated man-hours.
7. The estimated material costs.
8. The total estimated cost of the work.
9. The date of issue to the shops.
10. The date the work order was returned and action completed by the shop(s).
11. A remarks section.

The work reception clerk, upon receipt of the request, records the work description, date of receipt, the name of the requestor and assigns a work identification number. The clerk dates the document and passes it to the maintenance supervisor.

C. The supervisor screens the work and assigns a priority code to the work.

D. Planning, estimating and engineering: The work is planned, estimated and engineered in accordance with the governing criteria and procedures. The entire package is then returned to the supervisor for review, inspection and approval of the estimate.

E. The work reception clerk then prepares a maintenance work order for issue to the maintenance shop(s). A work order is not prepared until the request has been approved as required.

F. The supervisor reviews the work order to assure that it is complete and comprehensive, authorizes work accomplishment and approves it for scheduling. One copy of each pertinent sketch is sent to engineering to reflect changes, if any, for "as-built" drawings.

G. The scheduler takes the necessary action to obtain the materials and equipment required and backlogs or schedules the work as appropriate.

H. The scheduler, in accordance with the priority code, schedules the date for beginning the work and the date for completion. The scheduled beginning date will be a Monday and the scheduled completion date will be Friday. Example: Scheduled start date - Monday, February 12. Scheduled completion date - Friday, February 23. If more than one shop is involved, a lead shop will be indicated on the work order.

I. The shop supervisor schedules the accomplishment of the work for the shop within the time frame of the schedule. If more than one shop is involved in the work order, the lead shop supervisor coordinates the work with the other shop(s). Upon completion of the work, the completed, signed and dated work order is returned to the work reception center. If the work order cannot be completed by the scheduled completion date, the shop supervisor will submit the reason in writing, with a new completion date, to the work reception center. The work reception clerk then assigns the new completion date.

J. Upon receiving the completed, dated and signed work order, the clerk closes the work order to any further charges.

III. Identification

A. Purpose

The purpose of a work order numbering system is to identify the type of work being accomplished, the department the work is performed for, the originator of the work, to provide a method of work request and work order reference and to effect positive control over the work in the shops. A properly developed work order identification system will, in addition to the above, provide accurate processing and accomplishment data essential to controlled maintenance management.

B. Scope
Every work order (whether minor, specific or standing) is identified by a work order number which is both for reference purposes and for classifying and accumulating man-hour/material/equipment charges.

C. Method
The work order number is assigned by the work reception center and consists of eight (8) digits combined in the following example.

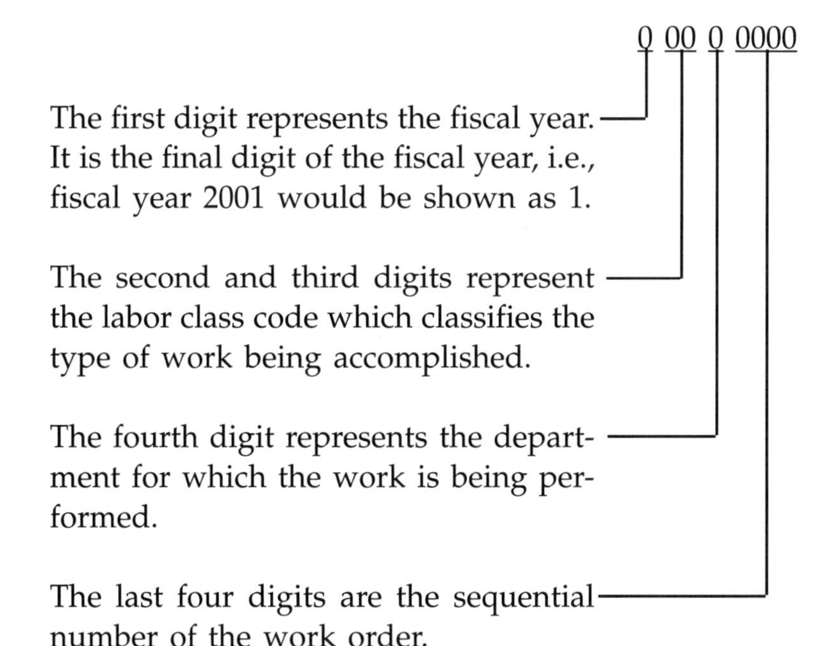

IV. Labor class and department codes

A. Purpose

Labor class codes are two digit figures that represent the various categories of productive and indirect/overhead work on which manpower and materials are expended. They are the means for gathering accurate data on manpower utilization within the maintenance department.

B. Structure

The suggested numerical designations for labor class codes are:

Productive:

01 Service work
02 Emergency work
03 Maintenance work - minor
04 Maintenance work - specific
05 Preventive maintenance inspections/service
06 Alteration or new construction - minor
07 Alteration or new construction - specific

Indirect/Overhead:

20 Rework
21 Shop supervision
22 Shop Indirect
23 Allowed time
24 General office supervision and clerical
25 Leave

C. Definitions of labor classes

The following clarification of the labor class codes is provided to assist in the determination and application of the proper labor class code to maintenance manpower expenditures.

1. Service work (labor class code 01) includes all productive work of a non-emergency nature that is minor in scope and does not exceed a predetermined dollar limitation.

2. Emergency work (labor class code 02) includes all productive work requiring immediate action to prevent loss or damage to property, to restore essential services or to eliminate an extreme hazard to health or personnel.

3. Preventive maintenance inspection (labor class code 05) includes all time expended in performing PMI.

4. Minor work (labor class codes 03 and 06) includes all productive work authorized by a minor work order.

5. Specific work (labor class codes 04 and 07) includes all productive work authorized by a specific work order.

6. Rework (labor class code 20) includes all labor used to correct faulty work.

7. Supervision (labor class code 21) includes all shop supervisory personnel and includes all time spent in supervisory duties by working foremen.

8. Shop Indirect (labor class code 22) includes personnel not chargeable to productive labor such as scheduler, planner/estimator, parts and material

issue personnel, labor engaged in the repair and maintenance of shop equipment and shop cleanup.

9. Allowed time (labor class code 23) includes non-productive time such as attending training classes, meetings, medical examination or attention, personal time and morning and afternoon breaks.

10. General office supervision and clerical (labor class code 24) includes plant engineer and other engineers, plant maintenance supervisor and clerical personnel assigned to the maintenance department.

11. Leave time (labor class code 25) includes time allowed for authorized and company paid leave, holidays and sick leave.

Note: Productive time is the time actually spent in performing assigned tasks. Time spent traveling to the job site from the shop, going after materials or parts and returning to the shop after completing an assigned task is NOT productive time.

D. Sample department codes

Department codes are used to designate the department for which the work is performed. Sample department codes are:

1. Administration
2. Finance
3. Production
4. Maintenance

CHAPTER 7

PLANNING AND ESTIMATING

The planning and estimating responsibilities of the maintenance department comprise a major part of the functional controls of controlled maintenance management. The controls include screening requests, planning and estimating the work, reviewing the estimate and work plan, scheduling and activating the work order and keeping informed on the progress and cost of the work.

Planning and estimating of work comprises all of the following:

I. Screening

The screening action performed by planning and estimating can save a considerable amount of time, effort and money. Requests for estimates and/or work should be analyzed carefully and if they fail to meet any of the following criteria they should be either revised to conform or they can be deferred and returned to the originator for more detailed information or justification.

A. Scope

The scope of the work must be understandable and should be compatible with the type of work that the maintenance department is authorized to perform.

B. Necessity

Requests for work may violate established maintenance standards or policies of the plant. These requests may be in conflict with current instructions or the work may be a duplicate of, or in conflict with, other planned maintenance work.

C. Availability of Funds

When funds to accomplish the work are not immediately available, processing should be deferred pending receipt of funds. If availability of funds is indeterminate, processing should be curtailed and the request canceled.

II. Engineering support

For work that will require engineering drawings, prints, sketches and/or research, no planning and estimating should be performed until the engineering support has been obtained.

III. Planning

The prepared work plan should specify the work that is to be accomplished, what is needed to do the work and which shop(s) will do the work. This means that complete specifications will be provided, the several operations that make up the job and, in some instances, the elements that make up the operations will be described. The clarity, correctness and completeness of the work plan is very important when there is to be accurate estimating, effective material coordination and realistic scheduling and work accomplishment.

Adequacy of Specifications

A. Adequate specifications are an essential part of every work order. The test of adequacy is the degree of reliance that can be placed on the specifications to define a complete bill of materials, obtain the necessary tools, special equipment or personnel, schedule the work and activate manpower with a minimum of visits to the job site by Supervision.

B. Material
The selection of materials is based on current policy, maintenance standards, fund limitations and/or other data.

C. Completeness
The planner must be meticulous in his responsibility for stating clearly and accurately the detailed scope of the work to be accomplished including sketches or drawings, as appropriate. This is necessary to ensure that the work will be performed in accordance with the requestor's needs and other governing criteria.

D. Method of Accomplishment
Jobs often require several operations and each operation may have several elements. By listing the elements and operations in their proper sequence, the following benefits are derived:

1. Important elements of the job are less likely to be omitted.

2. Related time values can be used in estimating.

3. Interrelated interests of various crafts for each work order are shown.

E. Assignment of work
The planner will indicate the appropriate craft for each operation and element of work listed on the work order and, also, the lead craft responsible for overall job supervision and coordination with others.

IV. Estimating

One of the most important functions of the maintenance department is estimating. An estimate is the informed analysis of all known and probable elements of a proposed job and the resulting detailed forecast of manpower materials, costs and related requirements that will be needed to accomplish the work.

A. Purpose
The principle purposes of estimating are:

1. To provide a basis for approval, disapproval or deferment of proposed work.

2. To provide a basis for budget forecasting.

3. To provide data for shop planning and scheduling.

B. Types of Estimates
There are three basic types of estimates and each serves a particular need.

1. Preliminary Estimate
   It is probable that in some instances work for which estimates have been requested will not be authorized; therefore, on questionable projects, and to eliminate unnecessary work, only preliminary estimates should be made in the early stages of these projects. Preliminary estimates are relatively simple computations made on an overall basis using up-to-date unit cost information as a guide.

2. Rough Estimate
   This is an approximation of man-hour requirements only. It is used primarily for standing work orders.

3. Final Estimate
   This is the type of estimate in which all work operations and elements listed are analyzed and considered in detail. It should be the most accurate forecast that can be made, within a reasonable time, of the costs and man-hour/material requirements for a given work order. Final estimates should be prepared for all specific work orders.

C. Estimating Criteria
The following factors should be considered in preparing estimates.

1. *Travel time* is the time required for the round trips between the shop and the work site for each man each day he works on the job. Also, the time spent procuring material for the job.

2. *Work preparation* is the time required for preparation of the work site, receiving instructions from superiors and the layout of materials and equipment.

3. *Work performance* is the time required for the actual performance of the craft work that is required to complete the work order. This requires an analysis of each phase of the work listed in the work plan.

4. *Material requirements.* It is the responsibility of the planner/estimator to specify the types of materials that are to be used and to estimate the realistic cost of the materials. A draft maintenance bill of materials should be prepared and screened.

5. *Equipment costs* must be included when it is expected that specialized equipment will be required to accomplish the work. Charges should also be included for maintenance department equipment used to accomplish the work on the work order.

6. *Cleanup.* This factor includes the normal cleanup during the performance of the work and upon completion of the work order.

7. *Contingency.* A contingency may be included as an additional factor in an estimate when a strict financial limitation has been placed on the job or when the funds are other than maintenance and operations expenditures. Generally a contingency factor should not exceed 10%.

8. *Overhead*. This factor should be applied in accordance with current policy and procedures.

VI. Review of work estimate and plan

The impact of the work order on the shops is of such importance that the maintenance supervisor should carefully examine the completed estimate before approval. The review will include the following factors.

A. Completeness

The estimate is composed of many items such as specifications, work plan, materials and equipment. The review should assure that none of the items have been omitted.

B. Accuracy

Technical descriptions, material computations, words and figures used throughout the estimate should be checked for accuracy.

C. Clarity

The content of the estimate should be clear to all personnel who may be concerned with its further processing. Ambiguous, lengthy or involved statements should be rephrased and expressed clearly. Keep it simple.

D. Conformance with Policy

The content of the final estimate should be reviewed for conformance with established policy.

VI. Activating the Work Order

A. Approvals

All necessary approvals for the work must be obtained prior to preparation of the work order. This should be accomplished in accordance with the established approval limitations.

B. Issuance of the Work Order
After the necessary approvals have been secured, a work order is prepared from the estimate, processed and issued to the shop(s) for accomplishment.

VII. Records

Work request and work order record files should be maintained in the work reception center. This will provide a centralized and readily accessible location for estimate, work request and work order records. Disposition should be in accordance with established procedures.

Planner/Estimator

One planner/estimator can provide the work planning and estimating for 30-40 craft people. The planner/estimator should be experienced in the work that is planned and estimated. A large maintenance department will require one or more planner/estimators for each of the electrical, mechanical, structural and instrumentation shops. In a small maintenance department, the plant engineer and/or maintenance supervisor can perform the planner/estimator duties.

Scheduler

The scheduler is also the work reception center clerk. The scheduler schedules the authorized work orders to the individual shops in accordance with the work order priority and the individual shops' workload assuring that the shop is not overloaded with work. The work orders are delivered to the individual shops prior to the start of the new work week.

Shop Scheduling

The shop supervisor should maintain a log showing the work order numbers, priority, scheduled start and completion dates. The shop supervisor will schedule the daily work for the shop and determine the actual start date of each work order with experienced assurance that the work will be completed in accordance with the scheduled completion date. The supervisor will submit all completed work orders and a report of all uncompleted work orders with explanations of why they were not completed per schedule and a new completion date to the scheduler at the end of each week.

CHAPTER 8

MATERIAL COORDINATION

aterial coordination is the control applied to maintenance department material procurement, inventory, preparation, storage, staging, and delivery.

MATERIAL COORDINATOR

It may be necessary to establish a material coordinator position to achieve effective material coordination. The material coordinator will accomplish the following.

- Establish a centralized location for the control and coordination of all maintenance department materials and parts.

- Provide a centralized source of materials and parts references.

- Provide a centralized location for the material status records.

- Establish a centralized source of material and parts information such as nomenclature, unit costs, availability and source of resupply.

METHOD

All maintenance department material requirements fall within two basic categories: (1) materials required in support of a particular work order and (2) materials required for the support of maintenance functions and emergency, service and minor work authorizations.

I. Maintenance bill of materials

Purpose

The maintenance bill of materials is used for the procurement of materials required for minor and specific work orders.

Method

A draft bill of materials is prepared by the planner/estimator based on the requirements of the work order. All pertinent data is entered at this time. The stock number and/or manufacturer's part number and the unit price should be entered.

II. Materials

A. Storage

Items that are requisitioned for general stock should be stored in an appropriate location under the jurisdiction of, and in accordance with, the procedures of the maintenance department.

B. Staging

Materials required for a particular work order should be staged until the entire requirement has been received. If the work order is for a large job, such as a remodel of a portion of the facility, the materials should be staged at a location near the work site. This will reduce the time of the craftsmen to obtain their materials and decrease the time for completion.

Note: Staging. To place materials in a particular location and held for a designated work order.

CHAPTER 9

SCHEDULING

Scheduling is the method by which shop personnel are firmly committed to work orders sufficiently in advance of accomplishment to assure the timely coordination of personnel, material and equipment. This is necessary to achieve maximum efficiency of work performance.

GENERAL

Scheduling is a carefully prepared advance plan of action that has taken into consideration the availability of manpower, materials and equipment; the proper sequence of the crafts necessary to perform the tasks and the most economical force to be assigned to the various tasks making up the complete job. Work orders are generally scheduled in the order in which they are received, by priority or the requirements of the customer.

The exact order in which work orders are scheduled is contingent upon the need for the work in relation to the mission of the plant; the availability of manpower, material and work sites; firm starting dates that have been established or committed in the pre-authorization stage; and the seasonal characteristics of the work.

Effective scheduling, when properly performed, provides for the orderly and economical accomplishment of work and the orderly introduction of work into the various shops.

Processing Work Orders

All work orders should be processed in the following manner.

I. Determining Material Action

The planner/estimator should review the bill of material to determine if the materials in the quantities required are available from the supply stocks or if special procurement action may be required. This action is performed to provide a preliminary determination for a starting date of the work order. If it is determined that a procurement action will have to be accomplished, a realistic material requirement date will be entered on the bill of material for scheduling purposes. If procurement of materials is needed, the work order should be placed in "Suspense" until all materials are on-hand and ready for issue.

II. Determining Work Action

Upon determination that all required material is available, the maintenance supervisor should take the following action.

A. Minor work orders

1. Issue

   The minor work order is logged out to the shop in a minor work order log. The log can be any record which will show the basic information pertaining to the work order to include the work order number, a brief description of the work, the total estimated cost, the total man-hours and the scheduled dates of issue and completion.

2. Backlog
The minor work order is entered in the backlog record and filed in numerical order with the other work orders in the backlog file. The tentative scheduled starting date is entered in the backlog.

B. Specific Work Orders
Specific work orders are usually backlogged until they can be scheduled into the shop(s) for accomplishment. An exception to this rule would be urgent work that must be accomplished immediately or by a specified date. In this case, all efforts should be directed to obtaining the necessary materials so the work can be scheduled.

Issue and backlog action for specific work orders is as follows.

1. Issue
The work order is entered on the master schedule showing the hours allowed (by craft) and the craft sequence for work accomplishment. Work is not started at this time.

2. Backlog
The work order is entered in the backlog record and then filed in numerical order with the other work orders in the backlog file.

III. Scheduling the Work

All minor and specific work orders are scheduled according to the classification of the work, its urgency and nature. Scheduling is discussed below.

IV. Work Completion

When the work specified in the work order has been completed, the lead shop gathers all shop copies, dates the lead shop copy and returns all copies to the maintenance supervisor. The supervisor will ensure that the work has been accomplished in a satisfactory manner by assuring that all the job specifications have been completed, that the work site has been cleaned and excess material has been returned to supply. The supervisor then assembles all the completed copies of the work order, clears his records and closes the work order.

Master Schedule Methods

Master scheduling is the application of shop loading techniques at various component levels to ensure that the workload is balanced with the workforce and that an adequate backlog exists to assure a constant and steady flow of work to the shops.

I. Master Schedule

A. Purpose

A master schedule is the visual focal point for the maintenance supervisor and is a visual tool for both the maintenance supervisor and management. It provides a centrally located place where management can review the status of all specific work orders in progress.

B. Scope

The master schedule schedules all specific work orders and those minor work orders for which it has been determined that scheduling controls should be applied.

C. Method

The master schedule is prepared on a weekly basis for a period of four to six weeks in advance. The specific work orders scheduled will generally be taken directly from the backlog or from requests at the time of receipt, depending upon the nature and urgency of the work.

1. The schedule is prepared based upon manpower, material and equipment availability during the week or weeks concerned. The maintenance supervisor should assure that the materials and equipment required are available when the work starts.

2. Committing Work

   Of the total man-hours available for minor and specific work, the maintenance supervisor should commit 70%-75% of the man-hours to specific work orders. The 25%-30% allowed for minor work will provide a cushion to absorb urgent work or unforeseen problems occurring during the work week. In arriving at the percentage of labor committed to the two types of work, a labor analysis should be performed (see page 46).

3. Adjustments to the Master Schedule
   The master schedule is a firm commitment of the maintenance department to specific work orders in accordance with the planned schedule. Once the plan has been accepted and agreed upon, adherence to the schedule is mandatory. This will assure that the work progresses in the most economical manner. The master schedule should not be adjusted to meet minor changes in work conditions or actual performance deviations. It should be changed only when there is a major change in the scope of the work, when major delays occur or when the entire schedule is disrupted by emergency conditions.

4. Determining Manpower Availability
   The maintenance supervisor should determine the man power availability far enough in advance to allow accurate scheduling. To determine the manpower availability, the supervisor must know the current strength, any anticipated manpower changes and the total man-hours committed to overhead, emergency and service work, and standing work orders. The computed man-hours available for minor and specific work orders is then committed according to the percentage factor allowed for each type of work. The man-hour figure allowed for specific work should not be exceeded in scheduling the work orders.

D. Application

The master schedule reflects, in addition to data pertinent to the work order, the actual man-hours that were expended against the scheduled man-hours for the week and the overall job progress.

II. Shop Schedule

The shop schedule is for the individual shops to plan the accomplishment of all work except for emergency and service work. It provides the daily work assignment of shop personnel for a one week period.

A. Scope

The shop schedule schedules all work for accomplishment to include standing work orders, PMI, minor and specific work orders.

B. Method

The shop schedule reflects the daily assignments for a one week period. The schedule is based on the master schedule manpower commitments and the man-hours committed to standing work orders with the remaining man-hours allotted to other work requirements. Priority work scheduling for the shop schedule is as follows:

1. Standing work orders are a continuing firm commitment of shop man-hours. Personnel thus employed cannot be scheduled for other work.

2. Specific work orders as scheduled on the master schedule are a firm commitment of shop personnel. Unless there is a major change in the scope of

the work, a major delay or an emergency condition which disrupts the schedule, compliance with the schedule is mandatory.

3. Emergency and Service work is provided for by allowing sufficient man-hours uncommitted to other work. The man-hours allowed are based on historical data.

4. Minor work orders are scheduled based upon the man-hours available for this category of work. They are also used as fill-ins if additional work is needed for the shop personnel.

C. Application
The shop schedule will reflect, in addition to data pertinent to the work order, the actual man-hours expended against the scheduled man-hours for each day and the overall progress of the job by the shop. The daily assignment of personnel on the shop schedule will give the shop supervisor better control of his personnel and will also eliminate delays caused in assigning personnel to particular jobs.

Material Coordination

On work orders requiring the procurement of materials from outside sources, the maintenance supervisor should accomplish the following.

- Backlog the work order until the materials and manpower are available and it can be issued for accomplishment.

- Monitor the processing of the material requirements to the degree necessary to assure delivery at the time required.
- Upon delivery of the material, he should process the work order in accordance with his established schedule.

Reports

The following reports are essential elements of master scheduling as well as a source of accurate date for management.

- The work order backlog report provides a comprehensive listing of all authorized work orders that have been backlogged due to manpower, material or equipment requirements.
- The work status report provides accurate information pertaining to the status of scheduled work orders.
- The shop variance report is used to make any necessary adjustments or revisions to the master schedule. It is prepared weekly on all scheduled work in progress.

Management Reports

The following weekly reports of the previous week's maintenance department activities should be forwarded to the plant management no later than Tuesday of each week.

- Work order backlog report with the- reasons for backlogging each work order.

- Work status report will inform the plant management of work orders completed in the previous week and the status of work in progress. The report will also state the reasons for work orders not being completed as scheduled, and the new scheduled completion dates.

CHAPTER 10

WORK PERFORMANCE AND EVALUATION

Work performance and evaluation are measures of the effectiveness of controlled maintenance management in achieving the goal of efficient utilization of manpower, materials and equipment within the maintenance department.

Work performance and evaluation are comprised of various factors which include the inspection of work both in progress and after its completion, the evaluation of completed work in terms of manpower and material expended and appraisal of the work performed through the utilization and analysis of comprehensive management reports.

The three primary factors used in work performance and evaluation are.

I. Work Inspection

While the work is in progress, and soon after its completion, the work site is inspected to determine the efficiency of the personnel performing the work, the quality of the work and if the work is performed in accordance with the specifications provided. Although physical inspection cannot be accomplished for all work being performed, the inspection of all important jobs and certain other jobs that will provide a work performance indicator should be accomplished. An analysis should be made for the individual jobs and for the total of jobs to arrive at an overall work performance appraisal.

II. Evaluation of Completed Work

A comparison and evaluation between the estimated and the actual labor, material and equipment expenditures is accomplished to determine the variations and, if excessive, to determine the cause of the variation. Careful analysis of the estimated and actual expenditures provides a useful tool in locating and correcting deficiencies within the planning and estimating procedures.

III. Management Reports

Management reports provide the essential data required for the final evaluation of work performed. These data, when combined with information obtained from actual work inspection and cost evaluation, provide an accurate means to evaluate the work performance on individual work orders and the department.

APPLICATION

The benefits to be derived from work performance evaluation include:

I. Location of Deficiencies

Deficiencies may be found in:

A. Shop Performance
Shop personnel may not be working according to the plan or specifications. There may be too many personnel on the job thereby reducing efficiency in work accomplishment. Or it may be the result of poor work performance or work practices.

B. Shop Supervision
The job may not be properly coordinated or scheduled by the shop supervisor. Daily work assignments may not be prepared sufficiently in advance. Proper job supervision may be lacking. Materials and equipment may not be available when required because of poor planning.

C. Master Planning
The work order may have been released for accomplishment before the materials are available. The sequence of operations may not be scheduled properly. Important elements of the job may have been neglected requiring an additional expenditure of labor, materials and equipment. The actual scheduling of the work may not have been realistic.

D. Planning and Estimating
The planning of the work may not have been properly developed. The labor and material estimates may not have been worked out in sufficient detail.

II. Provide Evaluations

Accurate evaluations can be developed for determination of the following.

A. Manpower Utilization
   Is the overall department manpower utilization in accordance with established standards? Where are the excessive expenditures of labor and what is the cause? How effective is the planning of manpower utilization?

B. Cost Effectiveness
   What is the ratio of manpower to material and equipment expenditures? Are job estimates within the acceptable range of variances allowed? Is there a continuing reduction in areas of waste and inefficiency?

In answering these questions and in locating areas of deficiencies, work performance evaluation is an essential element of controlled maintenance management.

Backlog Management

Maintenance planning and scheduling are two activities that ensure the allocation of resources and the sequence in which they are performed so all work can be performed at the least cost.

Backlog management is at the heart of the planning and scheduling process. Backlog is defined as the amount of unscheduled work that remains to be performed. The backlog should be reviewed monthly for backlog size and age. Some work orders or work requests may no longer be valid because of a change of plans. The invalid work orders or work requests should be discarded.

Backlog Size

The backlog should not have more than four weeks of work unless the older work cannot be started until a given date for a special reason.

Chapter 11

Buildings and Grounds

The objective is to maintain buildings and grounds in an economical manner that will be consistent with their functional requirements, sound architectural and engineering practices and appearance. For purposes of maintenance standards, buildings are divided into the following basic components for which these standards are established.

1. Structural features (foundations,. structural frames, trusses and similar items);
2. Roofing;
3. Interior finishes (walls, partitions, ceilings and floors);
4. Painting (interior and exterior);
5. Windows;
6. Plumbing and fixtures;
7. Electrical systems and fixtures;
8. Kitchen equipment.

Maintenance Standards for Buildings

Facilities to be Used for More than 10 Years.

Maintenance shall include all services and materials required to keep buildings in such a state of preservation that they may be continuously utilized for their intended purposes.

Structural features shall be maintained for stability of the structure. Special attention shall be given to foundations, trusses and framing including such items as bolts, anchors, connectors and other fastening devices to assure that adequate maintenance is provided.

Interior and exterior finishes should be maintained to correct all defects or damage to keep the buildings in good operational condition.

Roofing and flashings shall be maintained to prevent water or moisture from entering the building and to eliminate the necessity for renewal for as long a period as is practical. When renewal is required, the choice of materials for reroofing should be selected from those that will provide at least 20 years of service.

Painting. The necessity for exterior and interior painting should be determined on the merits of each case, taking into consideration such factors as geographic location, climatic conditions, the degree of deterioration of the painted surfaces, the functional requirements of the building and appearance.

Windows. Check for broken glass, looseness, deteriorated caulking. Thermography will detect leakage of cooled or heated air around windows. Window air leakage will increase the cost of heating and cooling.

Plumbing systems should be maintained in proper operating and sanitary conditions. Substandard piping and materials, equipment worn, damaged or corroded should be replaced.

Electrical systems shall be maintained in a safe and reliable operating condition. Deteriorated, obsolete or substandard components shall be replaced. The system must comply with the current issue of the National Electric Code.

Kitchen equipment should be maintained in a safe, sanitary and proper operating condition. Equipment that is worn, damaged or is in an unsafe condition should be replaced. When rearrangement of the kitchen facilities is required, consideration should be given to improving the functional operation.

Suggested Maintenance Standards

1. Buildings to be used for more than 10 years should be maintained as necessary to assure their most economical and efficient usefulness for an indefinite period.

2. Buildings to be used from 3 to 10 years should be maintained in a manner consistent with the planned useful life of the building.

3. Buildings planned to be used for less than 3 years should be maintained at the minimum acceptable standard without jeopardizing the health and safety of personnel.

4. Maintenance should be planned to permit orderly and economical accomplishment.

5. If maintenance falls short of the suggested standards above, a list of the projects that are to be deferred should be maintained in sufficient detail to clearly describe the defect(s) and the estimated cost of repair.

The basic objective is to manage and maintain the buildings in a manner that will facilitate the operations conducted thereon and, at the same time, protect the real estate from depreciation.

Ground Structures

The objective in maintaining ground structures is to maintain them in an economical manner that will be consistent with their functional requirements and reasonable appearance.

Ground structures include sheds, storage tanks (except those used in connection with utilities systems), tunnels and underground installations, flag poles, towers, walls, fences and gates. Not included are outside utilities systems, reservoirs, petroleum and gaseous storage facilities, central heating and power plants.

Grounds

Definitions

Grounds are all areas not occupied by buildings, structures, pavements, ponds and railroads.

Improved grounds are those on which intensive development and maintenance are performed. This normally applies to areas which contain lawns, landscape plants, golf courses and similar areas.

Semi-improved grounds are those on which periodic maintenance is performed but of a lesser degree than on improved grounds.

Unimproved grounds are all other areas not included in the above categories.

Land Management Plan

A land management plan will show the areas of improved, semi-improved and unimproved grounds together with land usage and capabilities. The plan should outline the maintenance, conservation and improvement programs necessary to increase the value of the land and to prevent the waste and destruction of natural resources.

The plan should incorporate necessary fire prevention and fire suppression measures essential to the prevention of waste and destruction of natural resources.

Maintenance Standards

Improved grounds should be intensively maintained at a level that will protect the investments and to present a satisfactory appearance. The work should include, but not be limited to, periodic mowing, fertilization, runoff and erosion control, weed, insect and rodent control, plant disease control, pruning of trees and shrubs and removal of debris.

Semi-improved grounds should be maintained in a lesser degree than improved grounds but at a level necessary to prevent dust or erosion problems.

Grounds improvement work should be in accordance with a landscape development program. The program should include establishment of priorities for all development and improvement work.

Control Measures

Mowing schedules should be regulated by the amount of growth. Grassed areas should be mowed at a height consistent with their use.

Fertilization and seeding. Soil classification should be used to assist in the establishment and maintenance of turf and other ground cover. Grounds should be fertilized to the extent necessary to maintain satisfactory vegetative cover and to promote the growth of healthy plants.

Irrigation should be limited to areas where supplemental water is essential to support vegetative cover and principle landscape planting.

Drainage, runoff and erosion control. All storm drainage systems and contributing areas should be maintained to prevent

(a) the ponding of water on, or adjacent to, paved areas,
(b) soil erosion,
(c) the formation of breeding places for disease transmitting insects, and
(d) the recurring flooding and ponding of intensively used areas.

Pavements

The basic objective is to maintain pavements in an economical manner that will protect the company's investment, reduce hazards to life and property to a minimum and permit continuous use at the designed strength.

Pavements are classified as portland cement concrete (rigid), bituminous (flexible) and miscellaneous. The term "pavement" covers all surfaced, paved and stabilized (other than grass) areas such as roads, streets, walks and vehicle parking areas.

All materials used in maintenance, repair and rehabilitation of pavements, whether accomplished by contractor or in-house personnel, should be in accordance with specifications The work should be inspected for compliance with the specifications.

Pavement failures should be investigated to determine the cause of failure and to assure proper corrective action. Repairs should be made promptly to avoid further deterioration. Surface repairs on a defective base or subgrade should be avoided. Repairs, generally, should conform to the original construction in strength, texture and appearance. Small areas should not be strengthened beyond the strength of surrounding paved areas.

Complete preservation treatment of bituminous pavements should be applied when required to revive cracked or weathered surfaces or to replace surface material worn or displaced by traffic in order to seal the top course against permeation of moisture and liven a surface that shows signs of oxidation and weakness. Spot treatment of scattered small areas should be performed whenever disintegration appears imminent. Such action will prevent isolated surface failure and extend the period between complete preservative treatment of large areas.

Complete resurfacing of pavements to correct rough or irregular surfaces should be done only when deficiencies in smoothness cause excessive traffic impact harmful to pavements or hazards to vehicle operation.

Overloaded pavements that have deteriorated because of continuous overloading should receive only minimum repair work that is necessary to retain them in operating condition pending installation of strengthening overlays or complete reconstruction.

Snow and ice removal, including erection of snow fences and application of abrasives or chemicals for ice alleviation, should be performed to accomplish pedestrian and traffic safety.

Utility Plants

Utility plants are classified in accordance with the purposes for which they were designed and constructed. Four general classifications comprise electricity generating plants: steam turbine, steam engine, gas turbine and internal combustion engine (usually diesel or dual fuel) driven generators.

Water plants for the provision of water services include those necessary for the production, pumping and treatment of water. The plants are divided roughly into three categories: production of water from wells, springs or reservoirs; pumping plants and water treatment plants.

Maintenance Standards for Utility Plants

The maintenance standards for utility plants must be high to assure the integrity and reliability of the plants. Safety of the operators and the equipment is very important. The PMI and PdM inspections will help to assure the integrity and safety of the plants.

Utilities Distribution Systems

The objective is to maintain utilities distribution systems in a manner that will be consistent with operating requirements, good engineering practices and protection of life, health and property, and that will assure economical and efficient utilization of the supplied utilities.

Steam Distribution and Condensate Return Systems

Steam distribution systems are defined as the steam piping between the point of supply (boiler plant wall or meter when steam is purchased) and the pressure reducing station or point of entrance into the building or structure housing the equipment that uses the steam. The condensate return system starts at the building line and ends at the boiler plant wall. It includes dump traps, flash tanks and pumps used to return the condensate to the boiler plant. Interior steam piping is not included. Steam distribution systems are classified as underground and above ground systems. The basic components of the system are:

Steam piping
Condensate piping
Expansion joints, expansion loops and pipe anchors
Valves
Traps
Condensate pumps, receivers and flash tanks
Insulation and/or covering
Structural supports

Hot water distribution systems are defined as piping in which hot water is circulated between the wall of the heating plant and the service entrance (building wall) at the buildings or structures in which it is used. Interior hot water lines are not included. The basic components of the hot water system are:

Piping

Valves

Expansion joints

Drains and vents

Water distribution systems are defined as all supply mains with necessary appurtenances through which water is transported between the source (wall of the treatment plant or pumping station when the water is produced by the water user or meter when the water is purchased) and the service entrance (building line) at the building served or other point of utilization. Nonpotable water systems for irrigation, fire protection and sanitary purposes are included. The basic components of the system are:

Supply mains and service lines;

Valves, valve boxes and manholes;

Hydrants;

Exterior drinking fountains;

Meters and equipment for measurement and control;

Reservoirs, elevated storage tanks and standpipes;

Chlorinators, chemical feeders, small treatment devises not included in treatment plants and pumping stations;

All appurtenant equipment such as automatic controls and cathodic protection devices.

Water distribution systems shall be economically maintained in good condition to provide, at all times, an adequate supply of water to meet all domestic, fire fighting, irrigation and industrial requirements without excessive loss from leakage. Potable water systems shall be maintained in a manner that will prevent contamination from sources exterior to the system. Extensive replacement of defective systems must be in accordance with current local criteria for new construction.

Electric power transmission and distribution systems are defined as:

Overhead and underground transmission and distribution lines from generating plants, or delivery points, to all main service entrance switches at buildings including substations and accessories. Exterior lighting systems include street lighting, flood lighting, perimeter lighting and security-lighting.

Maintenance Standards

The degree of maintenance, repair and rehabilitation of utility systems shall be governed by known foreseeable usage and preventive maintenance inspection criteria.

Electric power transmission and distribution systems shall be maintained economically and in good operating condition to provide reliable service in compliance with the standards of the National Electric Code. Extensive replacement of defective systems shall be in accordance with the current codes.

Fire Protection Equipment

The fire protection equipment covered here includes:

Automatic sprinkler and waterflow transmission systems;

Automatic fire detection and alarm systems;

Manual fire alarm systems;

Fire reporting systems;

Local fire and evacuation alarm notification systems;

Installed carbon dioxide fire extinguishing systems;

Miscellaneous installed fire extinguishing systems;

Standpipe and hose systems;

Water supply, pumping and distribution systems;

Hand-held fire extinguishers;

The above list does not include mobile and portable fire fighting apparatus and appliances primarily associated with organized fire departments.

Maintenance Standards

Fire protection equipment, including the essential components, should be maintained in full operating condition at all times. The emergency operating nature of the equipment must be recognized. The maintenance procedures must assure detection of otherwise unknown conditions.

Non-operational items such as hangers, fasteners and supports must be maintained to assure structural stability and preclude deterioration.

CHAPTER 12

PREDICTIVE MAINTENANCE

Predictive maintenance is more of an art than a science. It requires extensive training and years of experience to become an expert. Predictive maintenance is part of proactive maintenance. It means locating deficiencies or small problems in equipment and correcting them before they can become major problems.

VIBRATION ANALYSIS

Vibration analysis is one type of predictive maintenance. It utilizes specialized equipment to take signatures of the vibrations of rotating machinery. The signatures may be likened to the electrocardiogram (ECG) of a person's heart. The heart specialist analyzes the ECG signatures and determines the problem in the heart. The vibration analyst analyzes the vibration signatures of the machine and determines the cause of the excess vibration. All machines have an inherent vibration but only the excess vibration is analyzed.

The excess vibration may be caused by a bad bearing, an unbalanced condition, misalignment of motor and machine shafts, misalignment of gears and numerous other causes. After the analyst has determined the cause of the excess vibration, the maintenance personnel can correct the problem. More vibration signatures should be taken after the problem has been corrected to assure that the problem has been corrected.

If the excess vibration is minor, the analyst may recommend that nothing be done at that time but that it should be checked periodically. A chart of successive analysis signatures is made. The chart will probably show a gradual increase in vibration levels. This will allow the maintenance supervisor to predict when the component will fail and schedule the repair before a breakdown can occur. This is predictive maintenance.

Vibration analysts, with the requisite years of experience, can take the examinations for Certified Vibration Analyst I, II, or III.

THERMOGRAPHY

Thermography is the inspection of electrical equipment with an infrared camera. The service is performed on transformers, switches, motor control centers, motors and other electrical equipment. The analyst studies the photos and compares the item with a like item at normal temperature or the item's temperature rise above the ambient temperature. If the item's temperature is too high, the analyst will recommend replacing the item.

If the temperature of the item is only slightly above normal, the analyst may recommend inspecting it periodically. The temperature readings can be charted. This will probably show a gradual increase in temperature. The maintenance supervisor can schedule the replacement of the item prior to its failure and avoid a fire or power outage.

Thermographers, with the requisite years of experience, can take the examinations for Certified Thermographer I or II.

LUBRICATION

Lubrication is an important part of proactive maintenance. Equipment must be lubricated or it won't work. But which lubricant should be used in each machine? There are specialized lubricants for various types of equipment and usage, for example, turbine bearings, aircraft, automobiles, etc. Lubricants used in the manufacturing industry cannot be used in food processing equipment. Hundreds of thousands of bearings have been ruined by improper lubrication at a cost of many millions of dollars.

The maintenance department should consult with the manufacturers of the equipment and the lubricant suppliers for the correct lubricants to use in the various items of equipment. Over-greasing can sometimes be more damaging than under-greasing. The maintenance department should obtain formal training for all lube personnel to assure that the equipment is properly lubricated.

Lubricant analysis will show the true condition of the lubricant, e.g., water in the oil, metal-wear particles, breakdown of the additives, acidity, etc. Also, when the lubricants should be changed. Some plants that have changed oil on a scheduled basis without knowing the condition of the oil have learned, when they used oil analysis services, that they had been throwing away good oil. With oil analysis, the period between oil changes was greatly extended.

Chapter 13

Cost Accounting

Cost accounting is an extremely important part of controlled maintenance management. Without cost accounting, you do not know how much maintenance costs. You must keep accurate records of the labor and material costs on each facility and equipment item.

Cost accounting is very important and necessary for developing the maintenance budget for the next fiscal year. The records will show the actual costs, labor and materials for the various functions for the present year-to-date. The costs can be extrapolated to the end of the fiscal year and extended for the next fiscal year. This method will allow you to make estimated budgets for up to three years in advance.

Accurate cost accounting is necessary for the maintenance department. Maintenance budgets should be established for each department, e.g., each production department, administration, accounting, etc., in the plant including the maintenance department. All work for each department is charged against the department's maintenance budget.

In developing the labor costs, you must start with the hourly pay rate. To this is added the payroll burden. The burden includes the cost of fringe benefits, e.g., vacation, paid holidays, sick leave, group insurance, liability insurance, workers compensation insurance, training, etc. This information can be obtained from the accounting department. Also included in the labor burden is the cost and payroll burden of the maintenance department office and supervisory personnel.

Maintenance is a team effort. Each team member must do his part for the team to be successful. The maintenance team can be likened to a professional baseball team. The management of a baseball team maintains detailed records on each player. If a player makes numerous errors, he prevents his team from winning and is replaced. The players are always striving to make themselves more proficient and more valuable to the team. The same should be true of the maintenance team.

CHAPTER 14

SAMPLE FORMS

MANPOWER AVAILABILITY AND WORK PLAN SUMMARY

Manpower availability information is essential for work input control. Information on projected manpower resources is required to develop the work schedules for proected shop loading. By using this form, a three-month picture, by month, of available man-hours can be projected. The summaries are intended to depict the manpower available and the proposed utilization of the manpower.

I. Development of the Summary

Enter the month and year in the space provided (the summary for April, May and June is prepared in March), Use one summary for each month.

II. Use the following to determine the available man-hours.

A. Determine the number of work days in the month.

B. Multiply the number of work days by the number of hours in the established work day.

C. The above total will provide the number of man-hours in the month for one person. Multiply this total by the total number of personnel in the maintenance department and enter the figure on the Available Man-hours line.

D. Use the appropriate shop code for each column.

III. Development of Indirect and overhead Man-hour Figures

Indirect and overhead man-hours, although necessary, are not classed as productive work for developing the manpower availability summary. Rework is not productive work because the work was paid for the first time the work was performed.

A. Rework

Rework includes all labor used to correct faulty work by the maintenance department. Enter estimated man-hours based on historical data, anticipated work and management actions.

B. Supervision

Supervision includes the plant engineer, other engineering personnel and all supervisory personnel. The plant engineer and other engineering personnel time is allocated proportionately to each shop. If there are working foremen, the estimated time the foremen will devote to supervision is included on the Supervision line. The balance of the foremen's time is included in the work plan summary. If there is only one supervisor, the time is divided among the shops.

C. Shop Indirect
This includes personnel time not chargeable to productive labor such as scheduling, planning, parts and material issue, personnel engaged in the repair and maintenance of shop equipment and shop clean up.

D. Allowed Time
Allowed time includes time such as attending training classes, safety meetings, other meetings, medical examinations or attention, waiting for material, personal time, morning and afternoon break periods.

E. General Office and Clerical
All general office and clerical personnel time in the maintenance department is allocated proportionately to the shops.

F. Leave
The leave line includes all authorized leave, e.g., vacation, holidays, sick leave, etc.

G. Total Indirect and Overhead Man-hours
Total the man-hours, rework through leave, for each column.

H. Total Productive Man-hours Available
Subtract the total indirect and overhead man-hours from the available man-hours to determine the total productive man-hours available for each shop.

IV. Development of the work Plan Summary

A. Service Work

Service work includes all productive work of a non-emergency nature that is minor in scope and does not exceed a predetermined dollar limitation. This figure is based on current trends and historical data.

B. Emergency Work

This includes all productive work requiring immediate action to prevent loss or damage to property, to restore essential services or to eliminate an extreme hazard to health and/or safety of personnel. This figure is based on current trends and/or historical data.

C. Preventive Maintenance Inspection (PMI)

This includes the man-hours to be expended in performing PMI and services. It is based on the PMI schedule and the estimated standing work orders issued for PMI accomplishment.

D. Standing Work Orders - Unestimated

This includes all productive work that is authorized on a standing work order which is not estimated or scheduled (does not include emergency or service work). This figure is based on historical data and anticipated changes.

E. Standing Work Orders - Estimated

This includes all productive work that is planned, estimated and scheduled but does not include emergency or service work, PMI or shop indirect. This figure is derived from the actual man-hour daily requirements for the month.

MANPOWER AVAILABILITY SUMMARY

Month:

Available Man-hours

Function/Shop Code								Total

Indirect and Overhead Man-hours

Rework								
Supervision								
Shop Indirect								
Allowed Time								
General Office and Clerical								
Leave								
Total Indirect and Overhead Man-hours								
Total Productive Man-hours Available								

Workplan Summary

Service Work								
Emergency Work								
Preventive Maintenance Inspection								
Standing Work Orders—Unestimated								
Standing Work Orders—Estimated								
Minor Work								
Specific Work Orders								
Total Productive Man-hours Planned								

F. Minor Work
Minor work includes all productive work authorized by minor work orders.

G. Specific Work Orders
This includes all productive work authorized by specific work orders.

H. Total Productive Man-hours Planned
This is the total planned number of man-hours for productive labor. The total productive man-hours planned should equal the total productive man-hours available.

V. Audit the Figures

All figures should be audited or checked for accuracy by verifying the totals in all columns and their relationship with each other. The use of the manpower availability summary will allow a balanced work load for the maintenance shops. If a shop is under loaded for the month, unestimated standing work orders or work orders from the backlog can be used to fill in. If a shop is over loaded, should temporary help or a contractor be brought in or can some of the work be postponed?

Equipment History

The purpose of the equipment history form is to provide a record of all pertinent information on an item of equipment or a facility. It is an important historical record of the item. The form assumes that an item of equipment was new when it was acquired. You may want to note if it was used or rebuilt equipment when it was acquired.

EQUIPMENT HISTORY

Property No.: ______________________ Description: ______________________

Mfg.: ______________________ Model: ______________________ S/N: ______________________

Acquisition Date: ______________ Acquisition Cost: $ __________ Depreciation Rate: __________

Utility Requirements: Air Water Power ______________ Volts AC DC Phase ______________

Location: ______________________ System: ______________ Sub-system: __________ Key Req.: Y N

Date	Repair	Labor Hours	Labor Cost	Parts Cost	Total Cost	To Date

Acquisition cost includes the cost of the item, any shipping charges and the cost of installation with the item ready for use.

The depreciation rate is usually set by the accounting department. The rate helps you to determine the monetary value of the item at any time.

Repair. Record the dates of repairs, the repairs made including parts replaced, the number of labor hours, labor and parts cost and total cost of the repair. The "To Date" column is the total cost of repairs to date.

Periodic review of the historical records will allow you to determine if a particular part has repeated failures so corrections can be made. An analysis of the record will assist you in determining if the item of equipment is nearing the end of its economic life and should be replaced or will it be more economical to rebuild it.

Inspection Checkoff Card

An inspection checkoff card is necessary for PMI to assure that each item of equipment, facility and system has been properly inspected. It also shows who made the inspection. By totaling the inspection hours, the amount of time required for PMI can be determined.

1. Property Number
   Each equipment or facility item or system should have a code number assigned to it for identification purposes. Examples of code numbers are:

   Pumps - P1, P2, etc.

   Boilers - B1, B2, etc.

2. Description
   A brief description should be included to assist the inspector to identify the item if the property number has been removed.

3. Location
   The location of the item is given to allow the inspector to plan a route and sequence of items for inspection. This will save time and many footsteps.

4. Check Points
   The various things the inspector should inspect on the item are listed on the back of the card and sequentially numbered. The numbers are shown separately by column under Check Points. The inspector enters the appropriate code mark in the column under the Check Point after he has completed the inspection for each Check Point. If an "X" or "XX" has been entered, whether required or accomplished, a deficiency report is made showing the action accomplished or required.

5. Inspection Hours
   After the PMI of the item is completed, the inspector enters the number of hours, to the nearest tenth of an hour, that he spent in performing the inspection of the item. This is necessary to determine the amount of time required for all inspections and is used as a tool in scheduling labor.

6. Factors Affecting Inspection
   Indicate if a key is needed to enter the area where the item is located. Indicate if any special tools or tools other than what the inspector normally carries are required. This information can save time.

   A hand-held computer will make the inspection task much easier.

CODE					
NI	NOT INSPECTED	X	ADJUSTMENT REQUIRED	XX	REPAIRS REQUIRED
√	SATISFACTORY	(X)	ADJUSTMENT ACCOMPLISHED	(XX)	REPAIRS ACCOMPLISHED
				—	NOT APPLICABLE

FACTORS AFFECTING INSPECTION
KEYS TOOLS

DATE INSPECTED	FREQ		CHECK POINTS 1	2	3	4	5	6	7	8	INSPECTOR'S INITIALS	INSPECTION HOURS STD.	INSPECTION HOURS USED	DEFICIENCY REPORT MADE YES	DEFICIENCY REPORT MADE NO
5-3	M		X	√	—	(X)	—	√	√	(XX)	J.S.		.3	√	
6–12			√	√	—	√	—	√	√	√	T.H.		.2		√

PROPERTY NO.	DESCRIPTION	LOCATION
P 18	Booster pump	N.W. corner Bldg. 2 (inside)

Front Side

Safety: Comply with all current safety precautions.

Check Points:

1. Wiring and Controls: loose connections, damaged insulation, loose or weak contact springs, worn or pitted contacts.
2. Pump Operation: failure to start, rough operation, inadequate suction, discharge and shut-off heads, drag, misalignment.
3. Packing Gland: excessive leakage.
4. Packing Gland (water-seal type): lack of or excessive leakage.
5. Anti-friction Bearings: inadequate lubrication
6. Motor: unusual noises, vibration, shaft end play, overheated bearings.
7. Supports and Piping System: unsound, evidence of misalignment.
8. Clean up: dirt and debris in area.

Back Side

Completed Work Order Report

The purpose of this report is to provide final cost data on completed work orders for work performance evaluation, utilization and analysis.

1. The Completed work order Report is prepared on a weekly, bi-weekly or monthly basis as locally determined.

2. The estimated man-hours and material/equipment figures are from the work order file and entered in the appropriate column. The actual man-hours and material/equipment cost are entered in the appropriate column after the work order has been completed and the variance between the estimated and actual costs has been computed. Compute the total variance for the work order and enter the figure in the variance column.

3. A summary of estimated and actual expenditures by the department can be prepared from the above information and entered at the end of the Report.

4. Application
   a. On work orders totaling less than $1,000 a variance of 10% or more should be investigated. Investigate only that portion of the work accomplished having a variance in excess of 10%.
   b. On work orders totaling more than $1,000 a variance of more than 5% should be investigated.
   c. On work orders totaling more than $10,000 a variance of more than 3% should be investigated.

COMPLETED WORK ORDER REPORT

Week Ending:

Work Order Number	Labor Hours		Labor Cost		Material Cost		Equipment Cost		Total Cost		Variance Plus or Minus
	Est.	Act	Est.	Actual	Est.	Actual	Est.	Actual	Est.	Actual	
20103176	26	27.5	208.00	220.00	22.00	20.00	—	—	230.00	240.00	+4.3%
20104177	6	5	48.00	40.00	4.50	4.50	8.00	9.50	60.50	54.00	–10.7%
20203180	2	1.5	16.00	12.00	28.00	24.50	—	—	44.00	36.50	–17.0%
Total	34	34	272.00	272.00	54.50	49.00	8.00	9.50	334.50	330.50	–1.2%

Above are examples of the typical breakdown of work orders.

5. Variance Investigation
   When an investigation into the variance is completed, it should reveal the cause and permit corrective action. The cause could have originated in estimating, planning, scheduling, material cost changes, work performance, or a combination of any or all of the above.

Labor Control Report

The purpose of the report is to provide management with information necessary to evaluate the utilization of the man-hours available within the department; to provide a means to measure the effectiveness of labor planning; and to provide relevant statistics for comparison with goals or accepted standards of manpower utilization. In addition, the report provides management with a detailed record of how the total man-hours available have been expended.

1. Method
   The report is compiled in two time phases;
   a. Planning Phase
      This is the planned usage of man-hours as established in the manpower availability and work plan summary report. The planned expenditures of man-hours are shown for line items 1 through 20 [columns (a) and (e)].

   b. Actual Man-hour Expenditure Phase
      This phase is accomplished as follows:

Plan — Planned
Vari —Variance

MAINTENANCE DEPARTMENT
Labor Control Report

Period Ending:

Control Element		Current Month				Year-to-Date				Acceptable
		Plan. (a)	Actual (b)	Vari. (c)	% Dist. (d)	Plan. (e)	Actual (f)	Vari. (g)	% Dist. (h)	Range
1.	Rework									0.3-0.6%
2.	Supervision									6-7.6%
3.	Shop Indirect									5-6%
4.	Allowed Time									2-3%
5.	General Office & Clerical									1.5-2.5%
6.	Leave									14-18%
7.										
8.										
	TOTAL									
9.	Indirect & Overhead Man-hours									28-32%
10.	Service									6-9%
11.	Emergency Work									1.5-2.5%
12.	Preventive Maint. Insp/Service									1.5-3%
13.	Standing Work Orders not Est.									
14.	Standing Work Orders Est.									
15.	Minor Work									
16.	Specific Work Orders									
17.										
18.										
19.	Total Productive Man-hours									68-72%
20.	Grand Total Man-hours									
21.	Productive Effort	%	%			%	%			60-72%
22.	Productive Man-hour Control	%	%			%	%			80-95%

(1) Take the actual man-hour expenditures for line items 1 through 20 [columns (b) and (f)] from the actual distribution of the labor during the previous month.

(2) The variance [columns (c) and (g)] is computed by obtaining the difference between the actual and the planned figures. It is shown as minus (–) for hours less than planned and plus (+) for more than planned.

(3) Year-to-date figures are computed by totaling previous reports to the current report.

(4) Percent distribution [columns (d) and (h)] is computed by dividing each line item of actual man-hours by the Grand Total Man-hours.

(5) Line item 21 (percent productive effort) is computed by dividing line item 19 by line item 20.

(6) Line item 22 (Productive Man-hour control) is computed by dividing the sum of hours for line items 12, 14, 15 and 16 by the total productive man-hours.

Deficiency Report

The deficiency report forms are carried by the PMI inspector when making the inspections. The inspector will fill in a deficiency report for each item of equipment, facility or system that has deficiencies. A description of the

DEFICIENCY REPORT

Sheet *1* of *6*

Property No.	Description	Location
P - 36	*Booster pump*	*Bldg. 43*

Inspector	Date

DESCRIPTION OF DEFICIENCY	Est. Manhours
Elect. wiring damaged. Needs to be	
replaced	*1.5*

deficiency(s) and the estimated man-hours for correction must be shown. For example

P2 Sump pump
Thrust bearing needs replacing 1 hour

Failure to give a description of the deficiency and the estimated man-hours for correction will result in additional time for planning and scheduling which will be wasted time.

Emergency/Service/Minor Work Authorization

The combination work authorization form is used to eliminate paperwork. The appropriate box (emergency/service/minor) is checked to indicate the class of work authorization. The other boxes are self-explanatory.

Maintenance and Support Work Order

A sample maintenance and support work order is accompanied by the drawings and a maintenance bill of materials.

FORM 1107

EMERGENCY/SERVICE/MINOR
WORK AUTHORIZATION

DATE AND HOUR CALL RECEIVED: *06-08-01* *0805*

STAND. W.O. NO.: *1-02-4-1967*

☒ EMERGENCY ☐ SERVICE ☐ MINOR

LOCATION	SHOP	REQUESTED BY
Bldg. 04	*02*	

JOB STARTED (DATE AND HR.)	JOB COMPLETED (DATE & HR.)	TOTAL M/H	MATERIAL COST	TOTAL COST
0815	*0915*	*1*	*0*	*$25.00*

DESCRIPTION OF WORK:

Water overflowing @ can wash

repaired and adjusted float valve

AUTHORIZED BY	SIGNATURE (PERSON PERFORMING WORK)
T. Jackson	*Pete Snow*

MAINTENANCE AND SUPPORT WORK ORDER	W.O. No. *1-10-06-1075* Date *06-11-01*

Building/property No. *B 1* Scheduled Completion Date *06-25-01*

Requested by *Tom Brown* Actual Completion Date *06-22-01*

Estimate: Labor $ *5,950* Material $ *10,835* Total $ *16,785*

WORK SPECIFICATIONS

Remodel offices — Rooms 216, 217, 218 per drawings.

Work to be done by contractor. Time & material

MATERIAL AND LABOR			LABOR HRS.	MATERIAL COST	
Lumber (doors, studs, paneling)				*7,125*	—
Carpeting				*1,650*	—
Electrical				*950*	—
Paint				*125*	—
Carpenter	@	*$4,128*	*172*		
Electrician	@	*900*	*30*		
Painter	@	*504*	*24*		
		$5,532			
		TOTAL	*226*	*9,850*	—
		TOTAL COST $		*15,382*	—

APPROVALS

A.T. Jones	*June 11, 2001*	*T. Brown*	*06-25-01*
SUPERVISOR, MAINTENANCE	DATE	MANAGER	DATE

Annual Inspection Summary

The annual inspection summary is used as a long range planning tool. It is used to plan funding for large projects or a number of smaller projects for a given year.

1. The summary is prepared in advance of the annual fiscal year budget meeting. It will list the project to be accomplished, the year it should be accomplished and the estimated cost.

2. The determination of which projects are to be entered is made by an analysis of the history files. The analysis will show a trend for nearly item of equipment. The maintenance supervisor is then able to project when major repairs or replacement is necessary. Facility items such as roof repair or painting are determined by inspection, judgment and the maintenance standard.

3. Use of the annual inspection summary combined with a projected work plan summary will allow you to project your work load and estimated budget requirements as far in advance as you wish with reasonable accuracy. The projections always include the cost of labor, overhead and supplies. It is usually not practical, however, to make the projections for more than five years.

ANNUAL INSPECTION SUMMARY

Prepared by: ______ Title		For year ending ______		
Submitted by: ______ Title		Sheet of		
Bldg. or Facility Number	Project	Year to be corrected and estimated cost		
		2002	2003	2004
206	Repair roof	7,000		
208	Repair roof	8,500		
578	Overhaul storm sewer lift pump No. 1	2,500		
586	Repair roof			25,000
597	Retube boiler No. 3	18,000		
612	Replace 2 sewer pumps		30,500	
	Replace 1,500 ft. of water main			30,000
698	Replace 2 refrigeration compressors		24,000	
701	Repair windows	1,050		
714	Paint interior	2,400		
715	Paint exterior and interior	14,000		
716	Paint exterior		3,500	
	Total	74,450	34,000	55,000

Chapter 15

Suggested Starting Preventive Maintenance Inspection Frequencies

Electrical

Code: M - Monthly
Q - Quarterly
SA - Semi-annually
A - Annually

Facility, Unit or Component	Frequency
Disconnecting switches	SA
Electrical grounds and grounding systems	SA
Electrical instruments	SA
Electrical potheads	Q
Electrical relays	Q
Lightning arresters	SA
Power transformers, energized	M
Safety fencing	Q
Cathodic protection system	SA

Fire Protection Systems

The standards of the National Fire Protection Association (NFPA) contain the criteria under which most fire protection systems in the United States of America are designed and installed.

NFPA 25-1995—Inspection, Testing and Maintenance of Waterbased Fire Protection Systems assembles detailed inspection, testing and maintenance recommendations from several existing standards. NFPA 25 does not include suppression systems using chemical and gaseous agents, portable fire extinguishers, exit safety components, alarm and detection systems, in-plant fire apparatus and other systems and equipment.

The suggested frequencies listed here are for water-based systems only. Contact your local fire department and your insurance carrier for review of your inspection system before implementing it. They may have requirements that you are unaware of.

The following information is reprinted from NFPA 25, *Inspection, Testing and Maintenance of Water-Based Fire Protection Systems* Copyright 1998. National Fire Protection Association, Quincy, MA 02269. This reprinted material is not the complete and official position of the National Fire Protection Association on the referenced subject which is represented only by the standard in its entirety.

FIRE PROTECTION SYSTEM

FREQUENCY

W = Weekly
M = Monthly
Q = Quarterly
S = Semiannually
A = Annually
3-5 = No. of years
VLC = Varies w/local requirements

ACTIVITY TYPE

I = Visual inspection
T = Testing
PM = Preventive maintenance

PERFORMED BY

B = Maintenance personnel
C = Contractor
D = B or C

See NFPA 25-1995

EQUIPMENT/TASK	FREQ	TYPE	BY
Control Valves			
Position (valves when sealed only)	W	I	B
Position (valves when locked or electrically supervised)	M	I	B
Tamper switches			
Position verification (physical)	Q	T	B
Full operation	A	T	D
Lubrication	A	PM	D

(Cont'd)

FIRE PROTECTION SYSTEM (*Cont'd*)

Fire Hydrants			
Condition and accessibility	W	I	B
Operability and water flow test	A	T	D
Lubrication and water drainage	A	PM	D
Fire Department Connections			
Condition and accessibility	M	I	B
Monitor Nozzles			
Condition and accessibility	S	I	B
Operability and water flow test	A	T	B
Lubrication	A	M	B
Main Line Strainers			
Internal component service	A	PM	D
Check Valves			
Internal component service	5	PM	D
Backflow Prevention Assemblies			
Isolation valves (unsupervised)	W	I	B
Isolation valves (electrically supervised)	M	I	B
Flow test	A	T	C
Internal component service	VLC	PM	C
Piping Systems			
Leaks, corrosion, damage	A	I	B
Fire flow test and friction loss analysis	5	T	C
Standpipe and Hose Systems			
Pressure regulating devices	Q	I	B
Piping			
Hose connections			

Standpipe and Hose Systems (cont'd)			
Waterflow and supervisory alarm systems	Q	T	B
Nozzles (operation) Hose storage devises (operation) Cabinets (condition)	A	I	B
Hose (unrack/unreel) Hose connections	A	PM	B
Hose hydrostatic test (see NFPA 1962)	3	T	C
Pressure regulating devises Water flow test Pipe hydrostatic test (dry pipe systems)	5	T	C
Fire Pumps			
Condition of pump house, pumps, motors engines, batteries, controllers and other components	W	I	B
Automatic pump start	W	T	B
Motor isolation switch and circuit breaker	M	T	B
Battery charging system Transfer switch (see NFPA 110)	M	PM	B
Engine fuel, breather, cooling and exhaust systems Battery terminals	Q	I	B
Engine flexible exhaust connection	S	I	B

(Cont'd)

FIRE PROTECTION SYSTEM (*Cont'd*)

Motor manual start	S	T	B
Engine coolant condition			
Motor electrical supply system	A	I	D
Engine exhaust and combustion air systems			
Pump performance test	A	T	C
Pressure and circulation relief valves			
Suction pressure control devices			
Alarm sensors and indicators			
Motor voltage and current			
Emergency manual starting			
Circuit breakers and fuses			
Overcurrent protection			
Fuel tank			

INDEX

SAMPLE FORMS